David Ginsburg

Implementing
ADSL

David Ginsburg

Implementing
ADSL

ADDISON-WESLEY

An imprint of Addison Wesley Longman, Inc.

Reading, Massachusetts • Harlow, England • Menlo Park, California
Berkeley, California • Don Mills, Ontario • Sydney
Bonn • Amsterdam • Tokyo • Mexico City

The publisher offers discounts on this book when ordered in quantity
for special sales. For more information, please contact:

AWL Direct Sales
Addison Wesley Longman, Inc.
One Jacob Way
Reading, Massachusetts 01867

Visit AW on the Web: www.awl.com/cseng/

ISBN 0-201-65760-0

Text printed on recycled and acid-free paper.

1 2 3 4 5 6 7 8 9 10-MA-0302010099

First printing, July 1999

To Jeanelle, who has helped me understand.
(There is beauty in truth, and truth in beauty :)

I have been to the end of the earth.
I have been to the end of the waters.
I have been to the end of the sky.
I have been to the end of the mountains.
I have found none that were not my friends.

Traditional Navaho Song

To laugh often and much; to win the respect of intelligent people and the affection of chil-
dren; to earn the appreciation of honest critics and endure the betrayal of false friends;
to appreciate beauty; to find the best in others; to leave the world a bit better, whether by
a healthy child, a garden, or a redeemed social condition; to know even one life has
breathed easier because you have lived. This is to have succeeded.

Ralph Waldo Emerson

Contents

Acknowledgments

Anthony Alles, Arthur Lin, Keerti Melkote, Gil Tene, Chin Yuan, and many others at Shasta Networks.

Craig Sharper, Sayuri Sharper, Enzo Signore, Rene Tio, Guy Fedorkow, and Tim McShane, all of Cisco Systems.

And finally, my current employer, Shasta Networks, and former employer, Cisco Systems, who provided me with the opportunity to research and complete this book.

Introduction

The growth of the Internet has been phenomenal. In just 15 short years, backbone capacity in the United States alone has grown from single T1s (1.544 Mbps) to multiples of OC-48 (2.4 Gbps), a 10,000 percent increase. In the same timeframe, applications have evolved from simple (and sometimes unreliable) email and FTP connectivity to the Web, e-commerce, Internet telephony, and video streaming.

However, until recently, access bandwidth has not kept pace. Except for dedicated corporate access, which has more or less followed the backbone bandwidth curve, bandwidth available to the typical residential user or even telecommuter has grown from 1200 bps to 56 Kbps, or in some cases ISDN at 128 Kbps, a paltry 1000 percent increase. Thus, a disconnect in the Internet's ability to support high-bandwidth content and the average user's ability to take advantage of it.

Only in the last year has this situation changed, with the introduction of the Digital Subscriber Line (DSL) and cable modems. These two technologies deliver, for the first time, megabit connectivity to the masses at an acceptable price point. This last point is critical, since nothing actually precluded running an OC-3 ATM connection to the house down the street: whether they could afford it was another matter.

This book focuses on DSL; specifically, ADSL, the flavor most relevant for casual Internet users and telecommuters. Irrespective of any rhetoric concerning provisioning and reach, ADSL is quickly growing in popularity, as evidenced by the number of users applying to local offerings and the ISPs signing on to support the service. Early difficulties in provisioning have been overcome, and both the ILECs and CLECs are now deploying the technology in quantity.

The availability of megabit connectivity to the average household opens up a wealth of new services, and enables true convergence between Internet content and media. For the first time, high-quality video streaming becomes a reality, and the "world wide wait" experienced when browsing graphic-intensive catalogs or auction sites is a thing of the past. For the telecommuter, ADSL delivers on the promise of productivity: access to corporate resources no different than from the corner cubicle.

ADSL is therefore an important technology to understand, in terms of both standardization, and more importantly, forming a basis for high-speed Internet services.

Who Should Read This Book

This book is for the networking professional who is either involved with ADSL deployment or is interested in how the technology plays a part in Internet service deployments and how the technology finds a role in practical network implementations.

This book covers a broad base of topics, including standardization, the hardware components found within a typical ADSL deployment, and the various higher-layer protocols and services riding above this infrastructure. It ties these various components together into a service of implementation examples.

However, *Implementing ADSL* is not intended to be an in-depth treatise on any one of the subjects, such as the details of ADSL encoding or ATM traffic management and signaling. For this, the reader is referred to more detailed books on these individual subjects.

How This Book is Organized

This book is organized in the following manner:

- Chapter 1, "An Introduction to ADSL," presents a business case for high-speed connectivity as well as providing an overview of ADSL standardization.
- Chapter 2, "ADSL Architecture," takes a layered approach to a typical ADSL deployment, beginning with ADSL as a modem technology. Using this as a basis, it introduces ATM as the encapsulation of choice, and then IP as the network layer protocol. Associated with IP are the various transport and service protocols found within a typical ADSL deployment.
- Chapter 3, "ADSL Infrastructure," takes a step back, introducing ADSL from the hardware perspective and describing the various CPE, DSLAM, splitter, ATM switching, routing, authentication, and provisioning systems found within a service provider's network.
- Chapter 4, "ADSL Services," integrates the earlier descriptions, combining the various protocols and hardware into currently deployed end-to-end service models. These models include the first model deployed, bridging, as well as the more current PPP and PPP over Ethernet.
- Chapter 5, "ADSL Implementation Examples," then integrates this into actual implementation examples based on commercially available hardware. It includes examples of router and aggregator configurations, as well as the necessary subscriber provision-

ing systems. Applications include residential access, telecommuting, small business connectivity, portals, media distribution, and DSL wholesaling.

- Chapter 6, "Alternatives to ADSL," introduces alternative technologies, including cable modems, wireless access, and other DSL variants such as IDSL, SDSL, HDSL, and VDSL.

- Finally, the Appendixes provide a general ADSL resource, including services and tariffs; compliance; relevant documents from the ADSL Forum, ATM Forum, IETF, and ITU-T; vendor and service provider web sites, and a glossary.

Chapter 1

ADSL History and Requirements

This chapter begins with a discussion of Internet growth, including the deployment of new applications and content resulting in a demand for increased bandwidth. It next looks at the role of the service providers in the new, competitive environment. This includes an outline of how *Asymmetric Digital Subscriber Line (ADSL)* infrastructure providers and their client ISPs may actually derive revenue from an ADSL service—the compelling business case, without which ADSL would never be deployed. Next, the various *Digital Subscriber Lines (DSLs),* technologies covered in greater detail within Chapter 6, are introduced. An introduction to ADSL history and the role of standards bodies outlines how the industry and technology have evolved. The chapter closes with an overview of current deployments and tariffs. This last section is of course evolving and should be considered a snapshot of rollouts as of mid-1999.

1.1 Growth of the Internet

The demand for increasing bandwidth is insatiable. As vendors roll out increasingly sophisticated software based on ever more powerful processors, as Web content grows in complexity, and as IP telephony and video streaming enter the mainstream, Internet users demand higher speed connectivity.

Telecommuters and homeworkers seek increased efficiency, and branch offices wish to deploy new network-centric services for subscribers but cannot afford bandwidth at traditional leased-line pricing models. This is coupled with a change in expectations as the Internet takes on an increasingly important role in everyday life. Examples of this include e-mail connectivity, commerce, and most recently, Internet telephony. This latter application, along with its more sophisticated kin, videoconferencing, lends a case for *permanent connectivity*. Here, a user is always reachable, able to receive incoming connections in addition to the traditional environment where only outgoing sessions are possible.

Thus began the search for a means of connectivity offering permanent, high-speed connectivity at an acceptable cost for both consumers and businesses. ADSL fulfills both of these requirements, offering permanent, high-speed connectivity to consumers, telecommuters, and businesses. When grouped into customer segments, four types of Internet users most likely to subscribe to DSL services are apparent. The first major group includes those requiring Internet access:

- Residential users operating small offices or using the service for e-mail exchanges or commerce.
- Businesses relying on Internet connectivity as a portal to the world.

The second major group is those requiring intranet connectivity:

- Employees who access their corporate networks at night or full-time.
- Businesses that connect to ISPs or branches over the service.

This demand for connectivity, however great, on its own is insufficient to guarantee the deployment of higher speed technologies at lower tariffs. The effect of ADSL on Internet connectivity and the perceived utility of its service are most evident when comparing a typical file download time for the different access technologies (Figure 1.1). Luckily, there are also external factors at work in the form of increased competition to the traditional service providers and the availability of a deployable technology.

FIGURE 1.1.

File Download Times

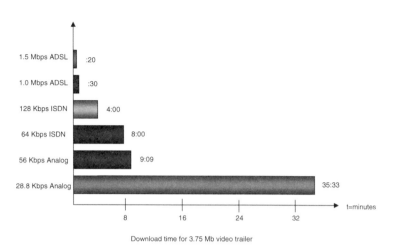

Download time for 3.75 Mb video trailer

1.2 Role of Service Providers

Service providers play a pivotal role in enabling this high-speed connectivity. These providers include the traditional telephony providers, known as *Incumbent Local Exchange Carriers (ILECs)* or *Regional Bell Operating Companies (RBOCs)* in the United States, and PTTs in many other countries. An incumbent carrier is the traditional, formerly monopoly or government-controlled, phone company in a given country that has built up the infrastructure over the last century. The incumbent carrier controls the copper in the local loop, as opposed to the newer competitive or alternative carriers that were originally formed to specialize in a given area, such as long distance services. These competitive providers are known as *Competitive Local Exchange Carriers (CLECs),* formed to compete with the ILECs in specific market segments in the United States and elsewhere. An even newer term is the *Packet CLEC (PCLEC),* which is a CLEC basing its infrastructure on packet switching technology. Finally, the *Internet Service Providers (ISPs)* are the Internet access providers, often making use of facilities provided by the ILECs/PTTs or CLECs. Where an ILEC or CLEC is providing the physical access infrastructure for an ISP, it is sometimes known as *Internet Access Providers (IAPs).*

All of these entities have a role in enabling ADSL service, and an understanding of the relationships between the incumbent and competitive carriers is instrumental in understanding how and where to deploy the different DSL technologies. It is also important to understand the role of regulation, which will determine the service offerings that may be deployed by a given class of subscriber. With these facts at hand, a business case may be developed that will demonstrate the viability of a typical ADSL deployment from the perspective of both the ISP and the ILEC.

1.2.1 Competition Among Service Providers

Until the birth of the CLECs, the ILECs and PTTs controlled both telephony service and any data services deployed across the local loop, be it copper or fiber. Access to the loop was tightly controlled, and tariffs for leased lines, depending upon the country, ranged from expensive to extortionary. The CLECs were the result of telecommunications deregulation in the United States and elsewhere. The Telecommunications Act of 1996 in the United States, for example, required the incumbents to unbundle their services and facilities. In this new environment, a CLEC could rent *dry copper* from an ILEC or locate some portion of their equipment within a central office (CO) in what is known as *co-location*. They then backhaul the subscriber traffic to their switched or routed core. More recent regulations within the European Community and elsewhere have had the same effect.

The growth of these new providers resulted in a more competitive service environment, with a net effect of reducing tariffs. Still, in 1998 the tariff for a typical T1 access link spanning 10 miles was not inexpensive. Fees ranged from $320 in the United States to over $1500 in those countries with less competition. These rates, although acceptable for organizations with large networking budgets, placed the service out of the range of those smaller users hoping to deploy applications previously confined to the LAN onto their corporate WANs.

As the new entrants in the market gained a foothold, the cost of backbone bandwidth began to drop dramatically. Two developments spurring this on were:

- The exponential growth of the Internet, which drove the demand for additional bandwidth.
- A technology known as *Dense Wavelength Division Multiplexing (DWDM),* which allowed providers to more cost effectively deploy high-bandwidth infrastructure by adding an extra layer of multiplexing based on colors. DWDM may be deployed on its own or in conjunction with SONET/SDH for greater bandwidth granularity.

A number of new CLECs such as Quest and Level3 were set up to capitalize on these developments, brokering bandwidth for both ILECs and ISPs. The local loop was the next step.

The two developments in this domain were the availability of a technology that would fit the ILECs business model, as well as competition from the CLECs/PCLECs and the cable providers. Understanding the ILEC business model is an important point, since existing technologies such as leased lines and HDSL, though successful, did not meet the needs of the mass market. This was due to their cost of deployment in contrast to what the average subscriber could pay on a monthly basis. In addition, leased lines and HDSL relied on the existence of additional cable pairs, acceptable for small businesses but not for many residential installations.

HDSL also introduces some binder crosstalk issues, precluding true mass deployment.

In the case of the CLECs, they too began to have access to local copper via the Telecommunications Act of 1996 and thus could deploy the same DSL technology as the ILECs. Meanwhile, the cable companies had a technology of their own, cable modems (discussed in Chapter 6), which became deployable in terms of cost and manageability about a year before ADSL.

Conventional wisdom holds that the ILECs began to deploy ADSL only under competitive pressure. Although this played a part, it was not the only factor, as the majority of the ILECs already understood competition before ADSL deployment. They also understand that they have a massive investment in a copper loop plant that is marginally utilized with the new technologies. This is a point-to-point infrastructure unlikely to be duplicated by the CLECs and unlike the cable providers' bus topology which requires a full network build-out to service the first customer. The advantage here lies with the ILECs that may now take advantage of their vast installed base in copper, an investment not easily replicated.

1.2.2 The Business Case for ADSL

Now looking at the ADSL business case, both the ILECs/PTTs and CLECs require some up-front analysis that a given service will actually prove popular. Predicting the ADSL market-size has been the subject of many surveys, both from independent analysts, the providers themselves, and the vendors. Opinions vary greatly, with analysts predicting from under a million to over 5 million global ADSL connections by the year 2001. At this high end is Ovum, predicting 19 million lines by 2003. Although these estimates may serve as a basis to generate potential business cases, and in fact are sufficient for that purpose, they very likely vastly underestimate the actual potential market-size. In addition, aggressive rollouts are only just beginning to occur in 1999. Witness SBCs ADSL tariffs or the Bell Atlantic / AOL announcement.

1.2.2.1 ADSL Market Projections

Looking at the growth in Internet connectivity, coupled with the increase in PC performance and multimedia content as described earlier, there is already an increasing demand for high-speed connectivity. This demand spans the continuum from telecommuters to casual Internet users. A more accurate DSL market projection should therefore be based on total Internet connectivity, then predicting the number of users served by high-speed media (both cable and DSL) in any given year. One assumption will be that this number will approach 100 percent of Internet users at some point in the future, say 20 years.

The next step will be to predict the percentage of such users served by DSL and the percentage served by other technologies. This is probably the most difficult step, since it will be based on how quickly the service providers and Multiple System Operators (MSOs) roll out their respective services, and how successfully they (or the ISPs with which they collaborate) market their services. Ultimately, in many locations, the average residence will be presented with two options, DSL and cable. One approach, depicted in Figure 1.2, is to state that 25 percent of Internet subscribers will have broadband connectivity five years from now, with 50 percent of these served by ADSL.

FIGURE 1.2.

Global ADSL Market Size

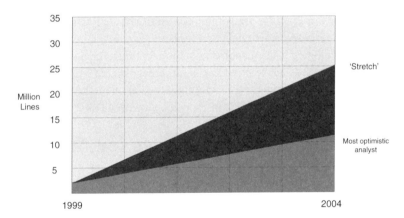

Using a very conservative estimate of 200 million Internet users by 2004 in the more connected countries, this still results in 25 million ADSL lines. Although this estimate may seem high, it still presents an idea of the potential market size for high-speed connectivity.

1.2.2.2 Deploying ADSL

Unlike the Internet, ADSL is not necessarily a "build it and they will come" type of service except for the more sophisticated users. Proper marketing of ADSL is therefore an important factor for success, since the ILECs in the United States and the traditional PTTs in other countries are not in most cases the dominant ISPs. These are usually other companies that have made this line of business their specialty, relying on the ILEC/PTT for the local loop. For example, Internet dial-in users would access the ISPs point of presence (PoP) via a local or long-distance phone call. This would, of course, leverage the switching and transmission infrastructure of the ILEC.

In the same way, the typical ADSL deployment includes an ILEC or CLEC providing the ADSL transport (and therefore operating the DSLAMs), and one or more ISPs providing the end-to-end Layer 3 service. Thus cooperation between the transport provider and the ISP is critical to ensure the success of the service. Two forms of cooperation are possible:

- Retail: The first is where the DSL provider and ISP operate as "ships in the night," with the former handing off ADSL ATM PVCs to the latter.
- Wholesale: Probably a more desirable relationship is where the DSL provider jointly markets the service with the ISP, a more desirable relationship. Here, the subscriber will purchase service (and even CPE) from the ISP. The ISP in turn will contract with the DSL provider for the local loop. A special case is the ISPs associated with the ILECs, such as BellSouth.net. Here, cobranding of the ADSL+Internet service can be very powerful, but the ISP must be very careful to uphold brand quality.

1.2.3 Analysis of an ADSL Business Case

In 1998, Cisco Systems in cooperation with Telechoice attempted to quantify the business opportunities for DSL deployment. The goal was to equip the service providers with a tool in which to baseline their deployment costs and expected return on investment. The analysis looks at the cost of ADSL deployment, as well as the expected expenses and revenues for both the ILECs building the ADSL and ATM infrastructure and the ISPs offering the end-to-end Internet service. This infrastructure, depicted in Figure 1.3, includes the ADSL CPE, Digital Subscriber Line Access Multiplexer (DSLAM), Service Aggregator, ISP router, as well as the necessary ATM infrastructure.

FIGURE 1.3.

Business Case Topology

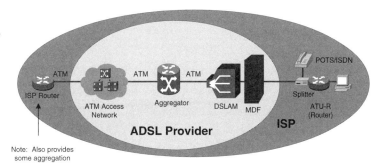

Systems controlled by the ILEC (the ADSL transport provider) include the DSLAM, aggregator, and the ATM access network, while the ISP is responsible for the CPE routers and the routers within its own backbone (which will also act as aggregators).

An alternative service model removes the aggregation function from the domain of the ADSL provider. This option is not covered in this book.

The two types of Internet services offered are residential and telecommuter. There are a number of different components that, together, comprise the total business model. In the following sections, these various components are detailed as part of an analysis of a flat-rate monthly service. The original business case details a usage-sensitive variant as well.

1.2.3.1 Estimating ADSL Service Demand

The first analysis calculates the expected demand for the ADSL service, data that drives the overall business model. Within the ADSL provider's service area, a realistic estimate of the number of COs served is added to the total number of loops. A percentage of these loops is ADSL-capable, while a smaller percentage still serve on-line customers who are considered ADSL candidates due to their high Internet usage. In selecting ADSL service, these customers will fragment into two groups: telecommuters who have their service paid for by their corporations and consumers who pay their own way. Table 1.1 summarizes this analysis for the first year of service. It provides an estimate as to the number of households within a provider's service area that are within reach of existing ADSL technology, and then goes on to predict the initial uptake. Note the two classes of subscribers—those funded by their corporations and those who purchase the service on their own.

TABLE 1.1. EXPECTED ADSL DEMAND

Number of central offices served	60
Total number of residential lines	3.6 million
% of lines ADSL-capable	75%
Number of ADSL-capable households	2.25 million
% of households with high Internet use	52% of 28%
Number of high-use households	327,600
% of these who take ADSL	17%
Number of ADSL customers at end of year	55,692
% of customers paid by corporation	19%
Number of corporate ADSL lines at end of year	10,693
% of customers paying for service	68%
Number of consumer ADSL lines at end of year	44,999

1.2.3.2 Estimating ADSL Revenue

The next step in the analysis is to look at the expected one-time and monthly revenues for both the the ISP and ADSL provider, as described in Tables 1.2 and 1.3, respectively.

Considerations here include revenue from ADSL line installation as well as that for the CPE. There is also a transfer cost from the ISP to the ADSL provider, in that it is the ISP that initially collects the payment from the subscriber. For example, in the analysis shown in Table 1.2, consumer service is $55 per month, collected by the ISP, and $35 of this is a transfer payment back to the ADSL provider. The ADSL provider will also collect a payment for ISP ATM access. This is detailed later in the analysis.

TABLE 1.2. EXPECTED ADSL REVENUES FOR ISP

Consumer ADSL Installation	$125
Consumer Monthly Tariff	$55
Telecommuter ADSL Installation	$125
Telecommuter Monthly Tariff	$80
Consumer CPE and installation	$199 + $150
Telecommuter CPE and installation	$500 + $300
ADSL installation + revenue to ISP	$26.9 million
CPE + installation to ISP	$22.2 million
Total ISP revenue	$49.2 million

TABLE 1.3. EXPECTED ADSL REVENUES FOR ADSL PROVIDER

Payment from ISP to ADSL provider per consumer installation	$63
Payment from ISP to ADSL provider per month for consumers	$35
Payment from ISP to ADSL provider per telecommuter installation	$63
Payment from ISP to ADSL provider per month for telecommuters	$50
Consumer + telecommuter installation to ADSL provider	$3.2 million
Consumer + telecommuter monthly to ADSL provider	$12.6 million
Revenue from ISP ATM access	$522,300
Total ADSL provider revenue	$16.3 million

1.2.3.3 Estimating ADSL Expenses

The expected revenue as detailed in Tables 1.2 and 1.3 are balanced by the cost of the service deployment, listed in Tables 1.4 and 1.5. From the ISP's perspective, the major expenses are the cost of the enhanced router infrastructure and the CPE. Major ADSL provider expenses include the DSLAMs, service aggregators, and ATM circuits.

TABLE 1.4. ADSL RELATED CAPITAL EXPENSES FOR ISP

Total CPE Cost (consumer and telecommuter) after discount	$9.29 million
Number of subscribers / router	1000
Number of routers	56
Price / router after discount	$21,700
Total router costs with overhead	$1.33 million
Total ISP capital expenses	$10.63 million

TABLE 1.5. ADSL RELATED CAPITAL EXPENSES FOR ADSL PROVIDER

DSLAM chassis cost	$9000
DSLAM modems / subscriber	$300
Subscribers / chassis	64
Number of chassis deployed	900
Total number of ADSL ports deployed	55,692
Total DSLAM Costs	$24.8 million
Service aggregation cost / subscriber	$20
Total service aggregation cost with overhead	$1.22 million
Annual cost per OC3 connection	$42,000
Annual cost per DS3 connection	$14,000
Subscribers / DSLAM	384
Number of DS3s from DSLAMs to service aggregators	180
Cost of connections from DSLAMs to service aggregators	$1.26 million
Number of OC3s from service aggregators to ATM network	8
Number of DS3s from service aggregators to ATM network	10
Cost of connections from service aggregators to ATM network	$266,000
Number of OC3s from ATM network to ISP	8
Number of DS3s from ATM network to ISP	10
Cost of connections from ATM network to ISP	$266,000
Total ADSL provider capital expenses	$30.7 million

The final section of the analysis relates to ongoing expenses for both the ISP (Table 1.6) and the ADSL provider (Table 1.7). These include marketing, operations, installation and provisioning, and ATM circuit costs. For the ISP, both provisioning and circuits are major expenses.

TABLE 1.6. ONGOING ADSL RELATED EXPENSES FOR ISP

Cost of sales and marketing	$3.2 million
Cost for customer service	$421,757
Cost of network operations	$275,037
Cost of on-site installations	$7.9 million
Provisioning and service installation	$9.3 million
Telecommuters per OC3 ATM connection	3000

TABLE 1.6. ONGOING ADSL RELATED EXPENSES FOR ISP *(CONTINUED)*

Consumers per OC3 ATM connection	6000
Telecommuters per DS3 ATM connection	1000
Consumers per DS3 ATM connection	2000
Monthly cost per OC3 ATM connection	$12,000
Monthly cost per DS3 ATM connection	$8,000
Monthly cost per PVC	$50
Number of OC3 ATM links to ADSL provider	8
Number of DS3 ATM links to ADSL provider	10
Total monthly cost for ATM links	$522,300
Total access costs	$16.9 million
Number of users per OC3 to Internet backbone	6000
Number of users per DS3 to Internet backbone	2000
Number of OC3s to backbone	4
Number of DS3s to backbone	2
Monthly cost per OC3 to backbone	$167,000
Monthly cost per DS3 to backbone	$53,000
Total backbone link costs	$5.9 million
Maintenance expenses	$277,760
Total ISP Ongoing Expenses	$36.4 million

TABLE 1.7. ONGOING ADSL RELATED EXPENSES FOR ADSL PROVIDER

Cost of line qualification	$619,784
Total sales and marketing (including line qualification)	$2.7 million
Total network operations cost	$849,875
Total loop provisioning and conditioning cost	$2.0 million
Total systems and tools costs	$535,000
Total DSLAM maintenance costs	$1.2 million
Total ADSL Provider Ongoing Expenses	$7.9 million

Provisioning from the standpoint of the ISP is the installation of the CPE and the activation of the end-to-end Internet service. The ADSL provider, meanwhile, is responsible for actual loop setup and conditioning (where required), which is recovered by the installation transfer payment from the ISP to the ADSL provider, described in Table 1.3.

1.2.3.4 ADSL Financial Outlook

Tables 1.1 through 1.7 detail the financials for Year 1 of the service, but it is also valuable to look at expected trends in Years 2 and 3. These are captured in Table 1.8, which clearly shows the expected pricing pressure in hardware costs, gains in installation efficiency, and resulting reduction in monthly tariffs.

TABLE 1.8. SERVICE TRENDS

	Year 1 Assumption		Year 2 Assumption		Year 3 Assumption	
	Consumer	Tele-commuter	Consumer	Tele-commuter	Consumer	Tele-commuter
Monthly price (Internet account & DSL service)	$55	$80	$55	$80	$50	$75
Service installation price (not including customer equipment)	$125	$125	$100	$100	$80	$80
Monthly price from ISP to DSL service provider	$35	$50	$35	$50	$32	$48
Service installation price from ISP to DSL service provider	$63	$63	$50	$50	$40	$40
Customer equipment price	$199	$500	$179	$450	$161	$405
Customer equipment installation price	$150	$300	$158	$315	$165	$331
Penetration rates for service	1.9%	1.9%	3.9%	3.9%	7.2%	7.2%

TABLE 1.8. SERVICE TRENDS (*CONTINUED*)

	Year 1 Assumption		Year 2 Assumption		Year 3 Assumption	
	Consumer	Tele-commuter	Consumer	Tele-commuter	Consumer	Tele-commuter
DSLAM capital cost per subscriber (CO side) including all DSLAM elements, tax, shipping, and test sets	$500	$500	$350	$350	$330	$330
Monthly price for DS3 / OC3 ATM connections	$8,000	$12,000	$8,000	$12,000	$8,000	$12,000
Monthly price for DS3 / OC3 connections to Tier 1 provider including transport and gateway fees	$53,000	$167,000	$53,000	$167,000	$53,000	$167,000

Note from Table 1.8 that ADSL equipment costs drop over time, but the cost of service and installation remains constant or grows. Thus there is no incentive to lower the monthly price of the ADSL service. As deployment becomes more widespread and Internet use in general increases, the penetration rate grows from 1.9% to 7.2%, increasing profitability. At least one provider has in fact stated that at this level of penetration profitability could be achieved much sooner. Finally, connectivity charges are assumed to remain constant over time. If these were to be reduced substantially, the ISP's business model would be enhanced.

Finally, Table 1.9 outlines the total revenues and expenses for both the ISP and the DSL access provider.

TABLE 1.9. OVERALL BUSINESS MODEL FOR YEARS 1–3

	DSL Summary		
	Year 1	Year 2	Year 3
ISP Revenues	$49,211,233	$169,301,334	$295,299,095
DSL Provider Revenues	$16,366,674	$66,456,390	$138,862,263

TABLE 1.9. OVERALL BUSINESS MODEL FOR YEARS 1–3 (CONTINUED)

DSL Summary			
	Year 1	Year 2	Year 3
Expenses:			
ISP Expenses			
Capital	$10,632,539	$29,115,536	$42,630,418
Sales and marketing	$3,233,835	$10,800,617	$18,913,958
Customer service	$421,757	$1,352,022	$2,361,891
Network operations	$275,037	$867,557	$1,507,403
Provisioning/installation	$9,319,122	$24,909,594	$25,191,094
Tools	$20,000	$4,000	$3,000
Access	$16,926,974	$68,360,190	$143,617,813
Backbone	$6,016,000	$23,776,000	$54,606,000
Maintenance	$277,760	$343,728	$425,866
Total ISP capital and expense	$47,123,024	$159,529,244	$289,257,443
Total ISP expenses (less capital)	$36,490,485	$130,413,708	$246,627,025
DSL Expenses			
Capital	$30,792,584	$62,949,599	$85,676,045
Sales and marketing	$2,719,784	$2,914,138	$1,998,371
Network operations	$849,875	$1,750,998	$1,524,522
Provisioning	$2,079,372	$4,084,226	$3,040,159
Systems and tools	$535,000	$26,750	$26,750
DSLAM maintenance	$1,240,380	$2,422,149	$3,154,132
Total DSL provider capital and expense	$38,216,994	$74,147,859	$95,419,979
ISP cash flow	$2,088,209	$9,772,090	$6,041,652
DSL provider cash flow	−$21,850,320	−$7,691,469	$43,442,284
Total cash flow	−$19,762,111	$2,080,621	$49,483,935
Discount rate for NPV	10%		
NPV of ISP cash flow	$14,803,012		
NPV of DSL provider cash flow	$6,427,391		
NPV of total cash flow	$21,230,404		
Years for straight line depreciation	5		
ISP depreciation	$2,126,508	$7,949,615	$16,475,699
ISP operating margin	$10,594,240	$30,938,011	$32,196,371

TABLE 1.9. OVERALL BUSINESS MODEL FOR YEARS 1–3 (*CONTINUED*)

DSL Summary			
	Year 1	Year 2	Year 3
Margin/revenue	22%	18%	11%
DSL provider depreciation	$6,158,517	$12,589,920	$17,135,209
DSL provider operating margin	$2,783,747	$42,668,210	$111,983,120
Margin/revenue	17%	64%	81%

The major cost elements for the ISP are:

- Equipment—If equipment costs were to be reduced, profits would increase to an extent. However, unlike the DSL provider, the ISP is more dependent upon core router technology where pricing may be less likely to change over time.

- Sales and marketing—Marketing expenses could be reduced, or at least amortized, over a greater number of subscribers by a higher penetration rate, better operations systems, and better use of techniques such as the web.

- Provisioning—With additional deployment, the cost of provisioning is also expected to drop, as the ISP gains experience with the service.

- Access/backbone infrastructure costs—Probably the greatest area for improvement is in the area of circuit costs, backing up the assertion made in the discussion of Table 1.8. If ISPs could reap the benefits of the lower cost of backbone connectivity deployment seen by some of the new carriers, their margins could improve substantially above those listed.

- Note that the ISP could also focus on offering additional value-added services to its subscriber base, allowing it to charge more on a monthly basis. These include VPNs, traffic management, filtering, and firewalling, and are described in Chapter 3 as elements of an ADSL infrastructure and in Chapter 4 as components of end-to-end services.

Now looking at the DSL provider, the major expense is for the ADSL hardware. As the providers are more successful in negotiating bulk discounts from vendors, this will be an area for improvement. In any case, as is evident by the margins depicted, the DSL provider enjoys a healthy business. Figure 1.4 depicts the cash flows and margin models for the business case summarized in Table 1.9.

What is obvious is that the ISP, although recovering from negative cash flow after the first year, will still have a tough time maintaining an acceptable margin since less of its expenses may be depreciated. This may be addressed by offering differentiated, value-added services resulting in higher revenue, or more important, working with an infrastructure provider with

FIGURE **1.4.**

Business Case Return
on Investment

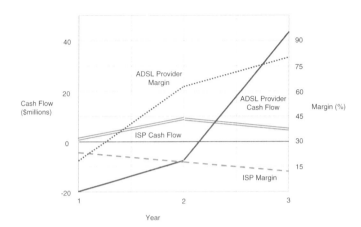

lower ADSL or ATM tariffs. In fact, the ISP capital expenses as outlined in Table 1.4 are a bit on the high side, but they pale in comparison to the price for access in Years 2 and 3. In contrast, the ADSL provider has a relatively robust business by the third year, based on both revenues from the ISP as well as reaping the benefits of capital depreciation since it has less ongoing expenses.

What is not obvious from the analysis is how the ILECs and ISPs can really derive revenue from the service. Although the business case is a good basic analysis of the cost of deploying the service versus the expected revenues, it just touches on the possibilities of a service-focused deployment. In the case of the ILEC, rather than providing a basic ATM VCC passthrough service to adjacent ISPs, it could offer Internet wholesaling and secure telecommuter access via tunneling or MPLS VPNs (both described in Chapter 4). Where desirable, the ILEC could also serve as a portal for high-quality video distribution via colocated ATM-connected video servers. As the regulatory environment evolves, ILECs will be able to offer additional value-added services such as VoIP with the promise of additional margin. Therefore, an awareness of what is critical in deploying a service-focused architecture is crucial, and focus on only the DSLAMs just won't do.

In the CLEC and ISP arena, the same influences apply. Either may offer wholesaling, and the delivery of value-added services or content by CLECs and ISPs (sometimes the same entity) is supported by the current regulatory environment in the United States. In all honesty, this actually provides the CLECs with a competitive advantage in some instances, given access to colocation and the local-loop. Elsewhere, the types of services permitted within the service portfolios of each of these entities varies on a country-by-country basis and evolves over time.

1.3 The DSLs

1

Although this book focuses on ADSL, an introduction to the different other DSLs is important. Previous to the DSLs, latent capacity within the copper cable plant went unused, as traditional voice traffic occupied only 0 to about 4 kHz, less than one percent of the available frequency spectrum. In cities and countries where ISDN is a factor, the copper loop is utilized up to 80 or 120 kHz, a 10 percent utilization. Still, the remaining spectrum up to about 1 Mhz was left unused due to the lack of sufficiently sophisticated encoding schemes. The following sections serve as an introduction to the various DSLs; Chapter 6 covers these technologies in greater deal, while contrasting them to alternatives such as cable, wireless, and more traditional means of connectivity including leased lines and Frame Relay. One point to note is that there are no standardized data models across these technologies, and in some cases, even the line encapsulation (such as ATM or Frame) is left to the vendor's implementation.

1.3.1 HDSL

The first technology to take advantage of this available spectrum was the High-Speed Digital Subscriber Line (HDSL), capable of carrying a T1 or E1 (1.5 or 2 Mbps) worth of traffic symmetrically over 2 copper pairs. A more recent form of the technology, known as HDSL2, requires only a single twisted pair. HDSL has seen acceptance by the incumbent service providers for delivery of leased line commercial services at T1 or E1 rates, or as a means of carrying multiple voice channels (24 or 30) between a serving CO and a remote terminal. This has proven quite useful, especially in increasing voice capacity for businesses or residential areas. HDSL is still a fairly expensive technology in comparison to ADSL, and until HDSL2, most vendor implementations were proprietary (meaning that hardware from a single vendor was required at both ends of the loop).

1.3.2 IDSL

The next DSL, the Integrated Digital Subscriber Line (IDSL), in effect reuses ISDN 2B1Q encoding but for permanent connectivity. Since it dispenses with ISDNs 16 kbps signaling channel (the D channel), its maximum data rate is 144 kbps symmetric over 1 copper pair. This bandwidth is suitable for telecommuters, where IDSL is primarily marketed. IDSL has been embraced by the CLECs, and to a lesser extent, the ILECs. A reason for acceptance by the CLECs is its use of 2B1Q encoding. In most locations, it is easier for a CLEC to receive permission to run this encoding over an ILECs local loops than is the case with alternatives (such as that used by ADSL). Thus the pain in commissioning a service is significantly less with IDSL than with ADSL. These encoding considerations are detailed in Chapter 2.

1.3.3 SDSL

The Symmetric Digital Subscriber Line (SDSL) delivers up to 768 kbps, also over a single twisted pair. As with IDSL, it uses 2B1Q encoding, making it appealing to the CLECs. SDSL has been a means of entry for CLECs hoping to take leased line business, and in fact has seen success in this space. It is cheaper to deploy than HDSL and, as noted, requires a single twisted pair unlike most HDSL variants that require two. None of these technologies, HDSL, IDSN, and SDSL, is capable of supporting traditional analog or digital (ISDN) voice traffic in the baseband. However, since HDSL and SDSL are aimed squarely at business users where a single additional voice channel would not add much value and IDSL is a telecommuter data service over second lines (many times replacing the analog modem traffic), this is not a major factor in actual deployment.

1.3.4 VDSL

The last of the DSL technologies, the Very High Speed Digital Subscriber Line (VDSL) is just beginning to see deployment in any quantity. Due to a distance limitation of approximately 3000 feet (1000 meters), it is suited to the DLC/FSAN environment as opposed to central office deployment. However, with this distance tradeoff comes a maximum data rate of up to 52 Mbps. Therefore, unlike the other DSLs, VDSL may be used to deliver one or more channels of high-quality video.

1.4 ADSL History and Standardization

ADSL, unlike the other DSLs, had its beginnings not as a technology for data delivery but for the transport of video services. Pre-1997 pilot deployments in fact focused on this application, with ADSL used to transport a digitally encoded (MPEG2) video channel from a video server to the subscriber directly over ATM without an internetworking layer. The control mechanism for this service varied from pilot to pilot, sometimes relying on proprietary protocols and other times based on IP. A common problem with these early deployments was the business case—whether a video-only service could be justified based on the technologies (server, settop, control) available at the time and whether subscribers would in fact partake of the service given the alternatives such as cable and direct satellite or video rental.

1

Unfortunately, the answer was no. This was coupled with the fact that high-quality video requires between 4 and 8 Mbps, effectively limiting the reach of the modems and, therefore, the percentage of subscribers served. These factors relegated ADSL to the sidelines for the time being.

Luckily, the explosive growth of the Internet provided the technology with a new outlet. ADSL, given some deployment difficulties that first needed to be overcome, was an ideal technology to deliver high-speed access that high-end subscribers began to demand. However, given Internet access as a baseline service, video distribution may prove viable in some instances, as described in Chapter 4. This is different from IP-based video streaming at MPEG1 quality and below, a major application within ADSL deployments. In addition, with the increasing deployment of ADSL within DLC deployments, some of the distance limitations of direct copper connectivity fall by the wayside.

The various standards that lie behind the hardware and protocols that comprise the end-to-end ADSL architecture actually fall under the jurisdiction of a number of organizations. These include the ADSL Forum, ATM Forum, the Internet Engineering Task Force (IETF), the International Telecommunications Union (ITU), and ANSI T1E1 to name the five major sources. Various components of the architecture may also draw on works from other bodies, described later.

1.4.1 ADSL Forum

The ADSL Forum was formed in 1994 and consists of over 275 telcos, semiconductor and equipment suppliers, system suppliers and integrators, and other interested organizations. This forum has the primary responsibility for defining architectural models, drawing on or adapting existing standards where required. Its first area of focus is the physical layer, including Transmission Convergence (TC) and Physical Media Dependent (PMD). TC includes ATM and Packet encapsulations, while PMD relates to the physical encoding. The ADSL Forum is not chartered actually to define any coding techniques (for instance, DMT). It is tasked with integrating them within an end-to-end deployable service architecture. In doing this, it must coordinate with other parties including the ITU, European Telecommunications Standardization Institute (ETSI), the American National Standards Institute (ANSI), the ATM Forum, the IETF, the Universal ADSL Working Group (UAWG), and the Digital Audio-Video Interactive Council (DAVIC) to name the major players. Current ADSL Forum efforts also include test and operations, as well as end-to-end service models.

The forum's output is a set of Technical References Recommendations. For example, TR-002, which describes the ADSL reference architecture when using ATM encapsulation, draws on the DMT standard from the ANSI and the ITU. The forum has also produced a set

of three basic recommendations (listed in the appendix) concerning ADSL and cross-refer-
encing ITU-T documents. These specify the use of DMT and/or UADSL as the modem
technology, along with the use of ATM as the encapsulation across the ADSL loop. Other
important documents resulting from the ADSL Forum include those dealing with service
architectures, management, and more recently, CPE.

The organization is composed of vendor and end-user representatives who submit contributions that are passed onto one of a number of technical working groups. These working
groups include:

ATM over ADSL (including transport   Testing and Interoperability
and end-to-end issues)
                                        Network Management
CPE/CO Configuration and Interfaces
Operations   Emerging DSLs (former VDSL)
                                        Study Group

For example, the ATM Transport sub-Working Group covers all issues pertaining to the
deployment of ATM encapsulation transport and transport interworking with ADSL across
the ADSL loop, and will liaise with the other groups when required.

A parallel Marketing Committee is the public relations component of the Forum, responsible for initiatives such as the Ambassador Program, summits, participation in trade shows
(for instance the ADSL Hot Spot at Interop, where vendors demonstrated interoperability),
and industry studies. The Marketing Committee is also responsible for the content of the
web site (www.adsl.com), which includes tutorials, meeting information, technical information, market studies, trial status, as well as a members' area for contributions, minutes,
and working texts. Membership categories include Principal Members, Small Company
Auditing Members, and Auditing Members, with the latter category of members not permitted to vote. The method of operation is very much in line with that of the ATM Forum, on
which the ADSL Forum structure is based.

1.4.2 Universal ADSL Working Group (UAWG)

Paralleling the efforts of the ADSL Forum, a number of service providers and vendors have
joined forces to promote a consumer ADSL variant known as G.lite. This lower-speed version of DMT is designed for splitterless operation. That is, instead of requiring an external
plain old telephone service (POTS) splitter, it relies on smaller in-line filters to protect the
POTS traffic from the ADSL data and vice versa. The stated advantage of this approach is
that it eliminates the costs associated with a "truck roll," where the ADSL provider must dis-
patch personnel to install the splitter. In Chapter 2 this technology is described in greater
detail.

The UAWG (www.uawg.org) was formed to promote the early adoption of G.lite. It consists of two groups of members:

- Promoters—deploy the technology or deliver the hardware and software to the consumer in support of G.lite. Vendors in this group include most ILECs, PTTs, Microsoft, Intel, and Compaq.
- Supporters—provide the DSL systems or chipsets. Vendors in this category include Cisco Systems, Alcatel, Diamond Lane, and others (listed in the appendix).

The UAWG had two areas of focus - transport and architecture. They developed a framework for presentation to ITU SG15 for G.lite determination (it became G.992.2) and having succeeded in their primary mission, have largely delegated completion of the work they started to other standards bodies (ADSLF, ITU, ATMF).

1.4.3 ATM Forum

The ATM Forum (www.atmforum.com) has responsibility for developing new or adapting existing ATM standards that may then be deployed by vendors and service providers. Originally, its charter was to adapt for the marketplace existing international standards such as those promulgated by the ITU-T. More recently, it has taken the lead in a number of areas where standards did not exist. The Forum has been active for over five years and has been very active in promoting the adoption of ATM by delivering a set of standards, which combined, provide the necessary functionality for public and private ATM deployment. Some of the better-known standards include signaling via the UNI 3.1 or Signaling 4.0, routing via the Private Network-Network Interface (PNNI), traffic management, the various physical layer interfaces, management, and security. Specific work is conducted by the various working groups of the Technical Committee. These include:

Physical Layer	Control Signaling/Routing and Addressing
Traffic Management	Service Aspects and Applications
Wireless	Multiprotocol over ATM and LAN Emulation
Residential Broadband	Network Management
Testing	Voice and Telephony over ATM
Security	

In the ADSL arena, the most relevant work is conducted by the Residential Broadband (RBB) working group. This group defines interfaces and service models for local-loop services based on the various DSL technologies and including video on demand (VoD). Of course, the various standards mentioned in the previous paragraph are also relevant in end-to-end ATM service delivery across ADSL.

In 1998, the ATM Forum also established a formal liaison to address issues pertaining to ATM over ADSL deployment. Specific areas of concern include the interaction of signaling, traffic management, and element management between the ADSL and ATM layers. This liaison is detailed in Chapter 2.

1.4.4 Internet Engineering Task Force (IETF)

Probably the best known of the standards bodies dealing with internetworking issues is the IETF. This organization, which has existed in one form or another from the earliest days of the Internet, has responsibility for all standards relating to IP, including data encapsulations, internetworking architectures, management, and related issues such as quality of service (QoS), multimedia support, and security. The IETF's Internetworking over NBMA Networks (ION) group has defined a number of architectures for the transport of data across ATM, many of which are relevant for ADSL deployment. These include protocols defining how both bridged and routed traffic is carried across an ATM network, along with mechanisms for user authentication and VPN establishment. These protocols are detailed in Chapter 2. Where required, the IETF draws on protocols from other bodies such as the ATM Forum. For example, the IETF protocol for signaling for IP traffic across ATM is based on the ATM Forum's UNI 3.1 or Signaling 4.0. In addition, the IETF's PPP over ATM RFC (2364) came from work started within the ADSL Forum.

1.4.5 International Telecommunications Union (ITU)

Of all the standardization bodies, the ITU is probably the most influential, especially with regard to traditional service providers and telecommunications vendors. It is an arm of the UN, with representation by national standards organizations such as ANSI in the United States. Responsibilities include a broad range of topics, and those most relevant to networking are covered with the Telecommunications Standards Sector (ITU-T). The ITU has developed many of the baseline ATM standards, later adapted by the ATM Forum. In the ADSL space, the ITU's most notable involvement is with the global standardization of ADSL through Study Group 15, Question 4. Areas of concern include DMT (G.992.1),

UADSL (G.992.2), start-up sequences (G.994.1), testing (G.996.1), and physical layer operations and maintenance (G.997.1). The appendix lists the various draft standards.

1.4.6 American National Standards Institute (ANSI)

ANSI T1E1.4 has been most instrumental in working to define the basic ADSL encapsulation standard in cooperation with vendors and service providers. In 1993 they sponsored an ADSL Olympics where the various encapsulation standards were evaluated and DMT was selected (over CAP) for further development and standardization. This has resulted in two issues of the DMT standard (T1.413), with a third issue planned. ANSI, as the U.S. representative to the ITU, then proposed DMT as an international standard, leading to DMT's current position as the global encapsulation of choice. Chapter 2 details the technical characteristics of both CAP and DMT.

1.4.7 European Telecommunications Standardization Institute (ETSI)

Parallel to ANSI's efforts in the United States, ETSI TM6 has developed an annex to the DMT T1.413 standard covering issues such as coexistence with ISDN and billing tones in the local loop. This is covered in the technical report ETR 328.

1.5 ADSL Deployment

Sadly, the deployment of ADSL on a global scale has seen numerous false starts and over-ambitious build plans described in press releases. Remembering that the technology was originally proposed for VoD, the first set of pre-1997 trials was very different in terms of architecture and business models than the later Internet-centric deployments. These first trials are not relevant to this discussion, in that global deployment was on the order of thousands of lines and based on an earlier generation of hardware. With the availability of stable and cost-effective ADSL hardware in 1997, service providers could begin to add ADSL into their business models, anticipating projected uptake and, therefore, revenues based on realistic tariffs.

The next step was to issue RFPs, promising national deployment of hundreds of thousands or even millions of lines in a short period of time. Most, if not all, of these ambitious plans did not come to fruition in the time predicted. However, by 1998 many ADSL trials (serving on the order of hundreds of users) were deemed successes, and by the second half of 1998 at least one provider, US West, had what could be described as a production deployment serving upwards of 10,000 subscribers.

The following sections detail deployments both within the United States and throughout the rest of the world, along with tariffs for announced services. Tariffing outside of the United States is still somewhat in flux due to less aggressive deployment schedules.

1.5.1 United States ILECs and the JPC

A major boost to ADSL in the United States was the announcement in 1997 of the Joint Procurement Contract (JPC) by three of the ILECs: Southwestern Bell, Ameritech, and Bell-South). This JPC was a statement by the ILECs that ADSL would become a major and universally deployed service, unlike the piecemeal ISDN implementations of a decade earlier. The JPC stipulated that Alcatel would be the sole supplier of DSLAMs and CPE for a set period of time unless new technologies appeared or individual customers requested a different DSLAM supplier. Bell Atlantic and US West did not participate in the JPC. The former originally chose Westell to supply DSLAMs, later shifting for volume deployment in 1999, while US West went with NetSpeed, a startup later acquired by Cisco Systems. GTE, the other major ILEC in the United States, announced the selection of Fujitsu in 1998 in preparation for commercial services. The impending mergers between SBC and Ameritech and between Bell Atlantic and GTE are bound to influence things further.

 NOTE The choice of an ADSL vendor implies just that—the DSLAM and the ADSL CPE modem. These two components are only a small part of the overall architecture described in Chapter 3.

Regionally focused ILECs such as Cincinnati Bell and Alltel also announced deployment plans. The larger IXCs were not silent, with MCI and Sprint both announcing services, along with a number of smaller CLECs such as Rhythms NetCommunications, Covad Communications, and Northpoint Communications . In all cases, the deployment plan would vary from region to region, with the CLEC either controlling the loop or leasing dry copper from the local ILEC. The service model in all cases was Internet connectivity, serving residential subscribers, telecommuters, and businesses.

Problems with trial deployments during the first half of 1998 included loop qualification, manageability, the cost of the truck roll in deploying the POTS splitter, and the requirement for deploying an end-to-end service architecture as opposed to only the DSLAMs and CPE. It was a learning phase for both the service providers and the vendors. Loop qualification was an especially sticky issue, due partially to the lack of comprehensive loop information in many cases, and due in part to the technical capabilities of the ADSL modems. Still, the technology received some negative publicity, sometimes along the lines of "my next door neighbor qualifies for ADSL but I don't."

1.5.2 Global Deployment

Deployments overseas paralleled that of the United States, with many incumbents announcing service plans and even vendor selections. However, none of the deployments except possibly that in Singapore equaled those in the United States. Reasons for this may have been due to lack of deployability (loop qualification problems) or a less pressing need to deliver high-speed services than in the United States. In any case, by mid-1998 actual deployments were running behind announcements, although a few of the providers had by that time announced more realistic rollout plans for the end of 1998 into 1999. Initial vendor selections included Westell, Orckit, Alcatel, Siemens, and Ericsson, although the choice of an equipment provider for a pilot did not necessarily imply that the same vendor would be selected for large-scale rollouts.

1.5.3 Tariffs

A major factor in whether ADSL succeeds is the tariff structure for the different classes of subscribers. For example, residential tariffs must be set in such a way to encourage "power-users" to transition from analog modems (and sometimes ISDN) to ADSL, while also remaining competitive with tariffs proposed by the cable vendors. For ADSL, it is a bit more complex since the basic ADSL service and the ISP component are usually operated by separate entities and may appear as two separate services on a monthly bill or may be combined. Telecommuters, less sensitive to tariffs than consumers, look for performance and dependability. Here, ADSL competes with ISDN and IDSL.

Finally, for business subscribers, the rates must be set to encourage uptake in the presence of leased lines, Frame Relay and SDSL. The complexity here is the possible cannibalization of existing high-revenue services. The major factor here is the level of competition by the cable providers or CLECs on a regional or national business. Where intense, it will tend to drive down tariffs. The often quoted statement "if we don't eat our young, our competitors will" is all too relevant here, where "young" refers to existing business services.

Appendix A outlines lines tariffs as published by many of the U.S. and international providers. Note the different types of offerings in terms of speeds, ISP bundling, and initial charges.

A Tariff Example

US West offered a number of options depending upon upstream and downstream bandwidth as well as whether ISP service was included. For example, the 512 Kbps symmetric "Mega Office" service, aimed at telecommuters, was priced at $62.40 per month for the ADSL component and an additional $22.55 for the ISP access. This tariff model was the norm in most regions, with the basic ADSL access decoupled from the ISP, allowing subscribers a choice of Internet providers. In addition to the monthly tariffs, the subscriber was faced with an initial installation and equipment cost, ranging from negligible if the provider wished to promote the service to hundreds of dollars in some cases. For US West subscribers, the set-up fee was $110, and the modem cost was either $199 or $299, depending upon the hardware deployed. This fee was waived if the user installed the modem.

One thing that these tariffs do not cover, however, is Service Level Agreements (SLAs) similar to those offered within traditional business leased line or Frame Relay tariffs. The very concept of an SLA has, of course, very little meaning, since the ADSL providers are at liberty to oversubscribe their services as much as they wish, in the same way that ISPs oversubscribe dial access. For casual residential ADSL access (tariffs of $59 per month) an oversubscription of 50 over or even 100 to 1 in terms of bandwidth may not be uncommon. In this case, 100 or more subscribers each at 384 kbps downstream may share a single T1 connection. Although this may not seem in the spirit of ADSL, it does allow providers to begin deployment of the technology at a cost per user on par with dial-up access, while bypassing the conventional switched phone network (the real goal). The use of SLAs is one way, in fact, that service providers will be able to continue to charge a premium for traditional leased line services. For example, they may set very quick response times for T1 outages, while an ADSL failure may take upwards of 24 hours to repair. For this reason, T1s should continue to remain popular among business users for mission-critical services.

1.6 Summary

There is a real need for the increased bandwidth at lower cost provided by ADSL. The time is ripe for deployment due to the current competitive environment, and both ADSL transport providers as well as ISPs may achieve acceptable returns on investment when offering the service. Due to new applications and more pervasive use of the Internet, demand will only grow. This, combined with a growth in the number of providers (both CLECs and ISPs) that encourages price competition, will further help generate demand. This introduction provides the groundwork for the more detailed analysis of the technologies and services in the chapters that follow.

Endnotes

1

ADSL Forum, ATM over ADSL Recommendations, ADSL Forum TR-002, March 1997.

Bell Atlantic, America Online and Bell Atlantic Form Strategic Partnership to Provide High-Speed Access for the AOL Service, January, 13, 1999.

Forrester Research, DSL's Field of Dreams, Telecom Strategies, April 1, 1997.

Megastream 2, British Telecom published tariffs, http://www.serviceview.bt.com/list/current/docs/Private_Cir_/01512.htm.

SBC Communications, SBC: Leader of the Bandwidth; California Offering Marks Biggest ADSL Rollout in Any State, January 12, 1999.

TeleChoice, DSL Business Case Template, http://www.cisco.com/warp/customer/779/servpro/solutions/dsl/telec bc.pdf, May 1998.

xDSL Press Release, Ovum Virtual Press Center, November, 13, 1997, www.ovum.com.

Chapter 2

Architecture

With an understanding of the business drivers and history leading to the current interest in ADSL gained from Chapter 1, this chapter delves into a discussion of the technology itself. It looks at the ADSL modem technology, focuses on the ATM encapsulation across the ADSL loop, and then describes the higher layer data models and protocols found within an end-to-end ADSL deployment.

One important point is in order. Although some service providers and many in the press describe ADSL as a "service," in reality it should be considered nothing more than a sophisticated modem technology, on par with V.32 or V.90 and the like. ADSL is in effect an encoding technology—over which may be deployed higher layer encapsulations like ATM, protocols, including IP, and higher order services such as Web access or multicasting—and, therefore, a layered approach is most suitable to describe the technology.

The use of layers is an easy way to conceptualize actions taken on data by different devices and at different points within the network. It also leads to a set of open interfaces on which developers may create necessary software and hardware. Within the internetworking community, operation has traditionally been divided into 7 layers (Figure 2.1), following the OSI reference model. Each layer provides services to the layers above and below it.

FIGURE **2.1.**

Layering Model

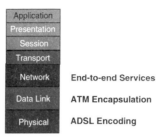

- Layer 1—The Physical Layer provides the physical infrastructure for all layers above it. ADSL and SDH/SONET are examples of physical layer technologies.
- Layer 2—The Link Layer provides controlled access to the physical layer. The best example of a link layer is Ethernet; however, in this discussion, ATM is considered a link layer as well, though this assertion has always been subject to interpretation.
- Layer 3—The Network Layer provides end-to-end connectivity through addressing and routing. The focus in the network layer is on IP.
- Layer 4—TCP and UDP are the two transports associated with IP.
- Layers 5 and 6—Session and presentation, respectively, are pretty much hidden from the end-user, as their functions are normally within a given application.
- Layer 7—The application is the one layer with which the user interacts.

Each of these seven layers wraps the original data, or packet, with header information necessary to perform its required function, and an internetworking device will operate only on those layers relevant to its function. Where a packet passes from one link layer technology to another, for example, the original header is removed and a new one added. However, higher layer headers are left untouched.

Within an ADSL deployment, the Physical Layer, described in Section 2.1, (Figure 2.2) handles the basic ADSL encoding, including the standards, data rates, and compatibility with other technologies in the copper loop. Standards here include Discrete Multitone (DMT), Carrierless Amplitude and Phase (CAP), and UADSL.

Layered over this encoding is a means of encapsulating higher layer protocols, associated with the Link Layer. Here, ATM, packet transport, and a third technology, Synchronous Transfer Mode (STM), described in the ADSL standards though not implemented, are the relevant options. Since the ADSL modems and DSLAMs deployed today rely on ATM as the encapsulation of choice, this is described in greater detail in terms of standards and architectures in Section 2.2. Closely associated with the ATM Layer is the ATM Adaptation

Layer (AAL), also described in 2.2. Here, data, voice, and video traffic is "adapted" for transport in ATM cells.

Section 2.4 introduces the various methods of transporting data across, referencing more detailed descriptions in Chapter 4. This of course includes voice and video in addition to data.

Finally, Section 2.5 describes higher layer protocols, with a focus on IP at the Network Layer and TCP as a transport layer protocol. As with ATM, IP is not the only internetworking protocol that will see use over ADSL; it is only the most predominant, especially in the context of delivering Internet connectivity. In any case, the section briefly describes IPX and the other Layer 3 protocols in use.

As part of this focus on IP, the chapter describes methods of tying together quality of service (QoS) and multicasting at the various layers in support of the services described in Chapter 4. As an aside, Frame mode operation is discussed, although this is far less popular.

FIGURE 2.2.

*Internetworking Encap-
sulations*

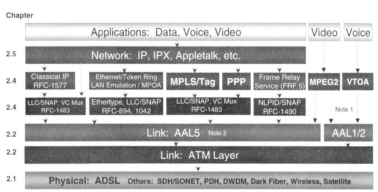

Note 1: MPEG2 into AAL1 possible but not standard within ATM Forum
Note 2: AAL3/4 also used for data transport in SMDS/CBDS but not relevant for ADSL

2.1 The ADSL Layer

This section provides a brief history of ADSL and introduces the various ADSL encoding techniques, focusing on CAP and DMT. Areas of emphasis will include not only the physical characteristics of the technologies by themselves, but also performance in the presence of "interferers" and, conversely, the effect of the technologies on existing services. A consumer-focused version of DMT known as G.lite is also described. This lower speed variant is promoted by the Universal ADSL Working Group (UAWG), which was introduced in Chapter 1.

ADSL is one of the most sophisticated of the various DSL technologies, capable of pushing mega-bits of data (along with traditional POTS traffic) down a single copper pair at distances in excess of 3–4 miles (5–7 km). It accomplishes this by implementing advanced encoding of the data, requiring processing at both ends of the copper loop. As with most new technologies, there has been a great deal of debate within the standardization bodies, this time focusing on the type of line encoding. In a repeat of the recently waged 56 kbps modem battles, proponents of the two competing ADSL encoding techniques, CAP and DMT, have promoted their merits in terms of per-formance, immunity to interference, complexity, and, ultimately, cost. Although both are capable technologies and one of the largest ADSL deployments to date is based on CAP (US West in the United States), it is becoming apparent that DMT will ultimately succeed. This is due to its favored status within the European Telecommunication Standards Institute (ETSI) and the International Telecommunications Union (ITU). From the standpoint of vendors, service providers, and consum-ers, it is not as much that one technology has won over the other; only that the dust has settled as ADSL moves from prototype to deployment. Luckily, the battle has mostly been waged within the halls of the ADSL vendor community as opposed to in the press as was the case with 56 Kbps. In fact, the debates over analog modem encoding may have slowed deployment to the extent that DSL and cable modems are now the favored option for infrastructure expansions.

2.1.1 ADSL Standardization History

ADSL standardization commenced in 1993 when the Regional Bell Operating Companies (RBOCs) evaluated three competing technologies: DMT, CAP, and Quadature Amplitude Modulation (QAM) at the ADSL Olympics, with a goal of selecting one technology for future work. Based on performance tests, they selected DMT, initiating the American National Standards Institute (ANSI) standardization process. This decision was later echoed by the ETSI.

Within ANSI, DMT is standardized as ANSI T1.413 (ANSI, 1998) and has been forwarded to the ITU as G.992.1 (ITU-T, 1998) and V.adsl . The DMT standard has gone through mul-tiple iterations, or *issues*, from the time ANSI selected it in 1993. Issue 1 completed in 1995 provides the basic framework, while Issue 2 provides better interoperability and includes references to ATM and rate adaption. For example, chipsets from multiple vendors, lacking interoperability at Issue 1, are capable of interoperation at Issue 2.

Worth noting are other parallel ADSL standards under development to complete the archi-tecture. Two of these are G.994.1 (ITU-T 1998a), which defines how the ADSL modem will negotiate parameters with the DSLAM, and G.996.1 (ITU-T 1998b) which defines test procedures.

Since the original decision in favor of DMT, ongoing debate within ANSI has focused on the relative merits of one technology versus another. By summer of 1998, the debate had cooled somewhat, with DMT progressing within the standardization bodies and less effort expended on furthering CAP. At the same time, almost every service provider had mandated DMT for new installations, lessening the impact of CAP. Beginning in 1999, most if not all new installations will be based on DMT.

During 1998, a further activity sponsored by a number of service providers, vendors, and system integrators (described in the appendix) founded the Universal ADSL (UADSL) Working group to speed up the adoption of ADSL in the residential market segment. The result of this effort is the G.992.2 standard, commonly referred to as G.lite or UADSL, a splitterless technology relying on the same DMT encoding as full-rate ADSL. Compliant DSLAMs in the central office will support both types of CPE—splitterless and full-rate.

Other goals of the standard include:

- Time-to-market
- Compatibility with POTS, fax, and V.34 modems
- Reuse of existing in-home wiring
- Avoidance of what is known as a *truck roll*, where service provider must visit the customer site for equipment installation, due to use of splitterless technology

The last point is quite important, in that it contrasts to most implementations of CAP and DMT that require POTS splitters at the customer, installed by telco service personnel. This installation involves a customer visit, costing upwards of $200, and directly influences the provider's revenue model In addition, only so many customers may be serviced in a given period due to the labor involved, slowing deployment.

However, with all of its advantages, there are some concerns regarding UADSL deployment in some countries, described later. At the CPE, modem compatibility between UADSL and analog technology is a major benefit, allowing consumers to deploy CPE capable of connecting to both dial and DSL Internet services. Although this type of CPE will not be capable of full-rate ADSL, it does remove a major barrier of entry to wide-scale deployment and will allow vendors to deploy a single chipset for multiple levels of connectivity.

2.1.2 Technology

The two technologies, CAP and DMT, are very different in how they encode the data across the local loop. Although both are frequency domain techniques, CAP relies more heavily on the time domain than does DMT, sending high bandwidth symbols across a wider spectrum for a short period of time. This is the *baudrate*, with CAP capable of rates up to 1088 Kbaud, or over a million symbols transistions across the link in a second.

In contrast, DMT relies on many smaller bandwidth channels, sending longer duration symbols at a narrower frequency. DMT's carriers are just over 4 kHz, capable of sustaining a bandwidth of about 32 kbps, close to that of an analog modem (which would be expected, since analog modems operate in the POTS frequency band which tops out at about 4 kHz). However, both accomplish their purpose in transmitting data (as was the case with VHS and Beta in recording video).

This section looks at the technical underpinnings of CAP and DMT in terms of modulation schemes, processing, performance, and interference. Both CAP and DMT rely on POTS or IDSN splitters to separate the analog or digital voice traffic from the ADSL data.

2.1.2.1 Carrierless Amplitude and Phase (CAP)

Carrierless Amplitude and Phase (CAP) is a DSL encoding technique relying on single downstream and upstream carriers occupying a larger proportion of the available bandwidth. Figure 2.3 depicts this spectrum, including the POTS traffic in the baseband. The technology segments the available spectrum into two single carriers, with the upstream between *f1* and *f2* and the downstream between *f3* and *f4*. The figure also depicts the Power Spectral Density (PSD) of the signals, with the upstream at –38 dBm/Hz and the downstream at –40.

FIGURE 2.3.

CAP Frequency
Spectrum

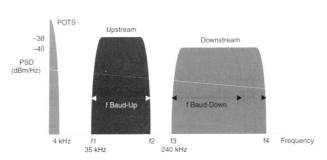

f2 and f4 depend upon baudrate as listed in Table 2-2

CAP modems are capable of accepting ATM, packet, and/or bit synchronous traffic, but as with most ADSL deployments, ATM will predominate across the loop. In addition, an Embedded Operations Channel (EOC) provides for monitoring and troubleshooting of the ADSL modems. Figure 2.4 depicts the data flow for a CAP transmission.

The user data and EOC are in turn fed into the transmission convergence sublayer, which is responsible for framing and Reed-Solomon encoding and interleaving (Forward Error Correction). The signal then passes to the physical media dependent element, which performs scrambling, trellis encoding, channel precoding, and the actual CAP transmission.

Figure 2.4.

*CAPFunctional
Diagram*

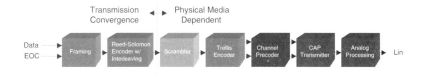

Note: Reed-Solomon is optional in upstream direction

Looking at the data traffic in more detail, CAP in fact defines two types of traffic. The first is Class A, transporting either a packet or cell-based data payload. This channel is not sensitive to latency. A Class B service, however, is designed for latency-sensitive traffic. It may be used to carry a bit synchronous channel, such as an embedded 160 kbps ISDN signal. Class B bypasses FEC and is optional. These two data channels combine with the EOC and are fed into the ADLS modems as depicted in Figure 2.5.

Figure 2.5.

CAP Data Types

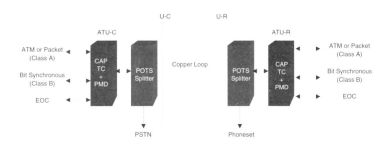

Class B, however, is in fact not expected to be deployed in the one major CAP installation, US West, so its fate is sealed. One reason for this is that CAP does not really need a separate path for low-latency traffic since its symbols are transmitted at a high baudrate. This contrasts to the relatively slow transmission of a DMT symbol, described in the next section.

2.1.2.1.1 CAP Baudrates

The CAP standard defines a number of downstream and upstream baudrates, or the number of symbols per second, as shown in Table 2.1. These are then used to derive the actual bandwidths based on the complexity of the constellation, as described in 2.1.2.1.2.

TABLE 2.1. CAP MANDATORY BAUDRATES

Downstream (kbaud ± 25 ppm)	Upstream (kbaud ± 25 ppm)
136	85
340 (required for startup)	136
680	
952	
1088	

In addition, the specification allows for a number of additional baudrates in accordance with the following rules:

- If N_0 and D_1 are integers in the range from 1–256, and D_0 is an integer in the range from 1–48, and $1/2.3 < N_0 / D_0 < 3.2$, then the downstream baud rate is ($34.56* N_0/ D_0 D_1$) in MHz.

- In the same way, if D_2 is an integer from 2–32, then the upstream rate is (2*downstream rate / D_2) in MHz.

Based on these calculations, a downstream rate from approximately 64–1088 kbaud and an upstream rate from 4–136 kbaud may be selected.

These baudrates result in the upstream and downstream frequencies as depicted in Table 2.2. They may then be applied against Figure 2.3.

TABLE 2.2. CAP FREQUENCY SPECTRUMS

	Downstream			Upstream	
Baud Rate	Start Freq f_3 (kHz)	Stop Freq f_4 (kHz)	Baud Rate	Start Freq f_1 (kHz)	Stop Freq f_2 (kHz)
136	240	396.4	85	35	132.75
340	240	631.0	136	35	191.4
680	240	1022.0			
952	240	1334.8			
1088	240	1491.2			

2.1.2.1.2 Constellations

The actual data rate is a function of the baudrate and the *constellation* size. The constellation is the number of encoded points per symbol and is no different than the encoding found within any dial-up modem. Constellations may vary in complexity from 8–256 symbols, reflecting increasingly dense data patterns, with the individual points relating to the amplitude of a sinewave (x axis) and cosine wave (y axis) as transmitted across the loop.

For example, an 8-point constellation (2 points in each quadrant) will be capable of encoding 6 bits per symbol (baud), while a 256-point constellation (the maximum currently implemented) will equate to 8 bits per symbol. Based on these values, a 1088 kbaud signal with a 256-point constellation yields the maximum CAP bit rate of 8704 kbps (1088 Kbaud * 8 bits/symbol). By the same calculation, the maximum upstream rate is 136 kbaud at the same constellation size yielding 1088 kbps.

2.1.2.1.3 Trellis Encoding

This maximum bandwidth is somewhat reduced by *trellis encoding*, a method of increasing the resiliency of the transmitted signal to noise across the line. As opposed to direct encoding of the data into one of the constellation points, trellis encoding introduces an additional step which gives the data an important property in terms of robustness. Figure 2.6 depicts the trellis encoding process.

FIGURE 2.6.

Trellis Encoding

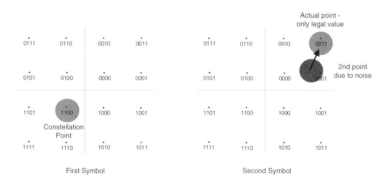

Given a symbol encoded to one constellation point (in this case 1100), the next symbol may only be encoded to one of a set of predefined values. If at the receiver, the point has drifted due to noise, it would then be matched to the one legal value out of a number of points in the area (or 0011 in Figure 2.6). This resilience is the reason why trellis is an option for the 256-point constellation within CAP—the large constellation increases the potential for error, which is reduced by the trellis encoding.

2.1.2.1.4 Reed-Solomon (RS) Encoding

Continuing the discussion of encoding, Reed-Solomon (RS), a form of Forward Error Correction (FEC), is used to provide additional resilience against line noise. First, RS codewords are formed. Next, these are passed through an interleaver (i.e., Ramsey Type-2) for increased protection against noise. The depth of this interleaver, D, is programmable and can be set to 1 (no interleaving), 2, 4, 8, or 16 code words. The interleaver's depths for the upstream and downstream channels can be set to independent values. Increased resiliency therefore comes at the price of additional latency.

2.1.2.1.5 CAP Frequency Spectrums

Referencing again the CAP frequency spectrum depicted in Figure 2.3, Table 2.2 lists the actual frequency range occupied at the various baudrates.

2.1.2.1.6 Quadrature Amplitude Modulation (QAM)

A second type of encoding defined by ANSI in 1997, though less widely implemented, is *Quadrature Amplitude Modulation (QAM)*. Currently, this encoding technique is included within the same document which defines CAP, although it is doubtful that the two techniques will converge. In contrast to CAP's Class A and Class B, QAM defines "Interleaved" and "Fast" with much the same objectives, and with terminology closer to that used within DMT. As Figure 2.7 depicts, interleaved data passes through FEC and is therefore more tolerant of latency, while fast data bypasses this stage and is therefore suitable for bit synchronous applications. As with CAP, QAM defines an EOC channel.

 NOTE

QAM is the encoding of choice for the downstream path in cable modems as well as an option for the upstream path. It is also the scheme used for the downstream direction within VDSL.

FIGURE 2.7.

QAM Data Types

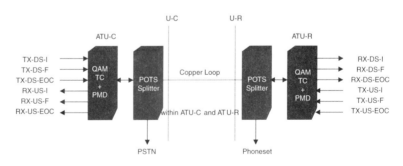

TX-DS-I	Transmit Downstream Interleaved
TX-DS-F	Transmit Downstream Fast
TX-DS-EOC	Transmit Downstream Embedded Operations Channel
RX-US-I	Receive Upstream Interleaved
RX-US-F	Receive Upstream Fast
RX-US-EOC	Receive Upstream Embedded Operations Channel

Note: Fast data bypasses Reed-Solomon encoder and interleaver

2.1.2.2 Discrete Multitone (DMT)

Although CAP was the encoding of choice for initial ADSL deployments, *Discrete Multitone* (DMT) is now the preferred method. DMT encodes the data into a number of narrow *subcarriers* transmitted at longer time intervals than CAP. As shown in Figure 2.8, DMT consists of 256 subcarriers spaced at 4.3125 kHz.

FIGURE 2.8.

DMT Frequency Spectrum

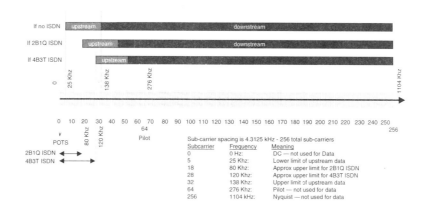

The modem may modulate each of these subcarriers at a different bit density (up to a maximum of 15 bits/sec/Hz or 60 kbps/4kHz tone) depending upon line noise. For example, at low frequencies where there is less interference, the line may support 10 bits/sec/Hz, while at higher frequencies this may drop to 4 at a corresponding decrease in bandwidth. In extreme cases, subcarriers are shut down due to interference. This use of subcarriers is one of the reasons that DMT is more complex than CAP in processing requirements, but it has been able to benefit by advances in DSP performance.

> DMT uses the same encoding as Europe's *Digital Audio Broadcast (DAB)*, where it is referred to as *Orthogonal Frequency Division Multiplexing (OFDM)*.

Within DMT, 0 kHz (DC) is unused, while the 256th carrier at the Nyquist frequency is not used for data. The lower limit for data traffic in the upstream direction (to the ATU-C) is determined by the POTS/ISDN filters. This also determines the upstream/downstream frequency split. Finally, a pilot tone modulated to (0,0) is carried within carrier 64 (276 kHz). DMT relies on an *Inverse Discrete Fourier Transform (IDFT)* for data modulation into each carrier, with the available bandwidth in each a function of the number of symbols. This results in a constellation size of varying complexity up to 256 points. Note that the UADSL variant of DMT uses only the first 128 subcarriers at a corresponding decrease in bandwidth.

2.1.2.2.1 DMT Operation

DMT supports both Synchronous Transfer Mode (STM) and Asynchronous Transfer Mode (ATM) data. In fact, STM was the only mode of operation specified in Issue 1 of the DMT specification. However, since that time, the cell-based ATM option has emerged as the encapsulation of choice, with STM bit synchronous transport the lesser known and unimplemented option.

STM defines a set of high- and low-speed bearer channels (Figure 2.9) that connect from the CPE to the ATU-R at the U-R interface and from the network core to the ATU-C at the U-C interface. The high-speed simplex bearer channels from the network core into the ATU-C (toward the user) may include AS0, AS1, AS2, and AS3, while low-speed duplex channels include LS0, LS1, and LS2. However, only AS0 and LS0 are required. All channels will conform to G.703/G.709.

FIGURE 2.9.

*DMT Functional
Interfaces for STM*

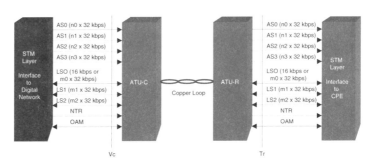

In ATM mode (Figure 2.10), a primary ATM cell stream (ATM0) conforming to I.361/I.432 passes through the cell transmission convergence stage and appears at the ATU-R as AS0 Optionally, a second ATM cell stream (ATM1) will map into AS1. The system also supports a Network Timing Reference (NTR) channel, defined with DMT Issue 2. As an example, in support of VTOA, this channel could support an 8 kHz timing marker derived from a Primary Reference Source (PRS) clock. The last channel between the ATU-C and ATU-R is the Embedded Operations Channel (EOC) used for both in-service and out-of-service maintenance as well as allowing the ATU-C to retrieve some ATU-R status information. In the future, this channel could be used for more sophisticated monitoring at the CPE. As with any ATM interface, the system inserts idle cells for cell rate decoupling, implements payload scrambling, and performs the header error check IAW I.432.

FIGURE **2.10.**

*DMT Functional
Interfaces for ATM*

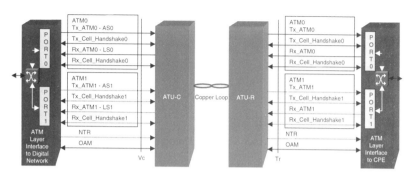

In the reverse direction, only the low-speed bearers (LS0, LS1, and LS2) are supported at the ATU-R, with only LS0, a requirement for STM and ATM. The case with ATM in the upstream direction is much the same as in the downstream, with LS0 a requirement and LS1 optional. As opposed to STM, these are the only subchannels really required since all multiplexing occurs at the ATM level via the VPI/VCI fields of the ATM cell header.

As with CAP/QAM, DMT defines two data paths: fast and interleaved. Fast offers low latency, while interleaved provides a low error rate via RSencoding at the expense of increased latency. At a minimum, a STM system will support both single and dual latency mode in the downstream direction, while only a single mode is required upstream. ATM requires only a single latency mode, with dual latency optional. Referencing ATM's ATM0 and ATM1, no fixed allocation of fast or interleaved data is assumed across the two channels.

The fast path mode of operation guarantees a maximum delay of 2 ms from the V-C to T-R reference points, suitable for the support of real-time traffic. In contrast, delay under interleaved mode is a function of both the Reed-Solomon codeword size and the interleave depth. The resulting maximum delay in ms is $(2 + S/4 + S \times D/4)$, where S ($=1, 2, 4, 8,$ or 16) is equal to the number of DMT symbols per R-S codeword and D ($=1, 2, 4, 8, 16, 32, 64$ — an integer multiple of S) is the interleave depth.

Mapping the various modes of operation (STM and ATM) and data paths into actual hardware, Figure 2.11 depicts a DMT ATU-C. ATM data first passes through an ATM cell transmission convergence function before entering the mux/sync control, while STM bypasses this initial step. The two data paths, Fast and Interleaved, are then split, both passing through a Cyclic Redundancy Check (CRC) function but Fast bypassing the interleaver. The data then passes into the physical media dependent section of the modem for tone and constellation encoding. This is where the 255 DMT subchannels are generated. Finally, the data enters the analog component of the modem for transmission across the ADSL loop.

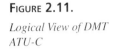

FIGURE 2.11.

Logical View of DMT ATU-C

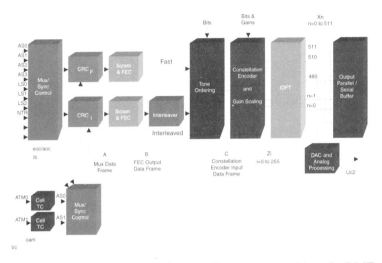

Now looking at bandwidth granularity, through the use of bearers mapped into the DMT subchannels, increments of 32 Kbps are possible. Referencing Table 2.3 and Figure 2.8, AS0 can support up to 192 bearers, yielding a line rate of 6144 Kbps if the distance allows. At system initialization, each bearer may be assigned a different latency path and mode of operation (ATM or STM). Proper choice of bearers will allow the system to support transparently a T1 or E1 service to the subscriber (noting that the system will not support T1 or E1 rates upstream).

TABLE 2.3.   DMT SUBCHANNEL BEARERS AND DATA RATES

Subchannel	Largest Multiple (x32 Kbps)	Highest Data Rate (Kbps)
AS0	192	6144
AS1	144	4608
AS2	96	3072
AS3	48	1536
LS0	20	640
LS1	20	640
LS2	20	640

Note: Lowest multiple is 1x32 Kbps except for LS0 at 16 Kbps.

2.1.2.2.2 Initiating Communication with DMT

When two DMT modems first wish to communicate, they must enter an initialization phase. Either end (the ATU-C or the ATU-R) may initiate this sequence:

- First looking at the ATU-C, after power-up or loss-of-signal, the device will transmit activation tones and will wait for a response from the ATU-R. If it makes two unsuccessful attempts, it will then wait for the ATU-R to initiate the connection. Alternatively, the network may request a retry.

- In the opposite direction, an ATU-R will repeatedly transmit an activation request after power-up. Upon receipt of an activation signal, it will continue onto the next phase of link bring-up. However, if it receives a C-TONE, it will cease activation requests for a minute.

- The next phase of link activation involves the two devices exchanging information regarding the throughput and reliability of the link.

- After transceiver training and channel analysis, they are ready to exchange detailed information regarding the number of bits and power levels to be used on each DMT subcarrier. At this point, the modems may exchange user data, a phase known as *showtime*.

Revisiting link-speed negotiation, DSL modems usually go through a lengthy start-up procedure which optimizes settings for line conditions and service requirements. Over time, however, these requirements may change. Instead of requiring the modems to renegotiate from scratch, DMT's Dynamic Rate Adaptation specifies a method whereby the modems continuously track line conditions. When changes actually occur, user traffic will be interrupted for a very short interval, on the order of tens of milliseconds. Another feature is the capability to reallocate bandwidth between the fast and interleaved paths when required. A variant of this technique, "Warm Restart," would allow modems to restart if the channel needs to be re-established.

2.1.3 Performance of CAP and DMT

The previous sections offer a good technical introduction to CAP and DMT encoding but it's important to realize that, from the standpoint of the user or service provider, actual performance in terms of throughput, distance, and reliability will determine the success of the service. The question always asked is: How far will this modem push my data at a given bandwidth (and over a given quality of copper)?

The answer to this simple question impacts the business model and the types of deployable services. As an example, a service provider may wish to offer a 4 Mbps ADSL service to business customers within range of a given CO. For each 1,000 feet of reduced distance, the customer base is reduced by the surface area. A technology offering 7,000 instead of 8,000 feet of reach would extend only to about 75% of the original surface area (38.46 vs 50.24), cutting off 25% of the subscribers. Thus any means possible of increasing bandwidth at a given distance is desirable.

Given that ADSL is based on a rather complex coding of data across the copper loop, its ultimate performance is highly dependent upon the throughput the loop is capable of sustaining at an acceptable error level. Although distance is the major factor, it also is influenced by some less visible characteristics such as the number of bridge taps, the gauge of the wire, other services in the same binder group, and even soil characteristics such as temperature and moisture content. This section reviews these dependencies and their effects upon performance.

2.1.3.1 Distance—The Shannon Capacity

The ultimate bandwidth is limited by what is known as the *Shannon Capacity* of the loop, a combination of the encoding scheme and the noise level across the loop. Figure 2.12 depicts this curve for CAP and the AMI encoding used within T1 carriers. As is obvious, ADSL encoding is a great improvement upon AMI, at least at intermediate distances.

FIGURE **2.12.**

Shannon Capacity

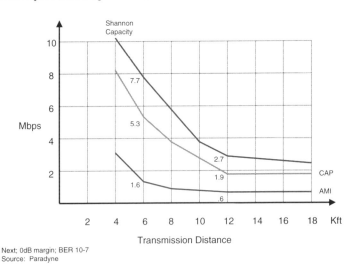

Next; 0dB margin; BER 10-7
Source: Paradyne

Tests determine first order performance for both CAP and DMT against a set of standard test loops that will determine the capabilities of the modems at different distances, over different wiring gauges, and in the presence of different interferers such as bridge taps. The *wire gauge* is the diameter of the copper, while a *bridge tap* is a point in the wire where a second, unterminated, wire is connected. Phone jacks within a residence are the best examples of this, though a cable serving multiple subscribers may be tapped as well. In the United States, ANSI T1.601 defines the relevant test loops (Figure 2.13) as well as those conforming to Carrier Service Area (CSA) rules (Figure 2.14).

FIGURE 2.13.

ANSI Test Loops

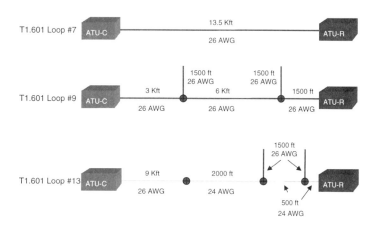

FIGURE 2.14.

CSA Test Loops

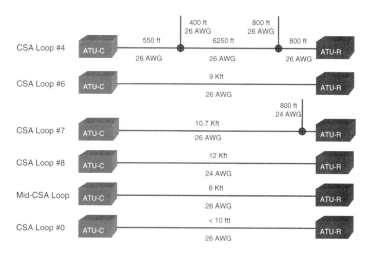

The CSA set of test loops is useful for assessing performance in the presence of Near-End Cross-Talk (NEXT) where two or more signals on adjacent pairs interfere with one another.

Within Europe, ETSI defined another set of test loops (Figure 2.15). In all three cases, the descriptions are pretty straightforward in their depiction of loop lengths, gauge (AWG in the

United States, and mm in Europe), and the number and length of bridge taps. Note that some of the test loops can in fact model complex topologies, including changes in cable diameter. Section 2.1.3.2 covers another set of measurements relating to the effect of ADSL on other services within the cable binder. These measurements are critical in determining a carrier's willingness to deploy the technology

FIGURE 2.15.

ETSI Test Loops

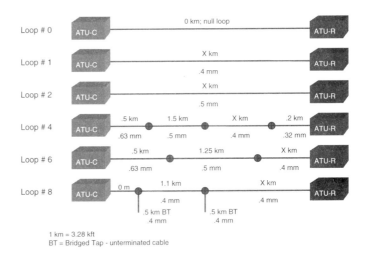

Now looking at Tables 2.4 and 2.5, ADSL modems are expected to support the listed bandwidths over these test loops with a maximum Bit Error Rate (BER) from crosstalk of 10^{-7} and a minimum noise margin of 6 dB. This 6 dB figure is the difference between the interference level at which the error threshold is crossed and 0 dB.

TABLE 2.4.   EXPECTED (DMT) ADSL PERFORMANCE OVER TEST LOOPS

Loop Sets	ATU Category	Net Data Rate			
		STM Only		ATM and STM	
		Simplex (AS0)	Duplex (LS0)	Downstream (AS0)	Upstream (LS0)
T1.601 (7, 13)	I	1536	16	1696	160
		1536	160		
CSA (4, 6, 7) Mid-CSA	I	5920	224	6144	224
T1.601 (7, 9, 13)	II	1536	16	1696	160
		1536	160		

TABLE 2.4. Expected (DMT) ADSL Performance over Test Loops (*Continued*)

Loop Sets	ATU Category	Net Data Rate			
		STM Only		ATM and STM	
		Simplex (AS0)	Duplex (LS0)	Downstream (AS0)	Upstream (LS0)
CSA (4, 6, 8) Mid-CSA	II	5504	640	6144	640
ETSI 1	I	2048	16	2208	160
		2048	160		
ETSI 2	I	2048	16	2048	32

The values in this table are only for testing—not for maximum distances

Category 1: basic—Trellis optional, EC or FDM mode

Category 2: optional—Trellis required, EC mode—and allows for overlapped spectrum

TABLE 2.5. Expected (CAP and QAM) ADSL Performance over Test Loops

Loop	Net Data Rate			
	CAP		QAM	
	Downstream	Upstream	Downstream	Upstream
T1.601 (7)	1700	816		
T1.601 (13)	2040[2]	816	2100[1]	750
CSA # 4	7616[1]	1088	6300[2]	1200
CSA # 6	7616[3]	680	6700[3]	750
CSA # 7	7616[4]	1088	6144[4]	1200
mid-CSA	7616[5]	1088	6144[5,7]	1200
mid-CSA	6528[6]	1088		

[1] In presence of 24 x DSL NEXT

[2] In presence of 24 x RADSL NEXT & FEXT and 24 x DSL NEXT

[3] In presence of 20 x HDSL NEXT

[4] In presence of 10 x RADSL NEXT & FEXT and 10 x DSL NEXT

[5] 4 x T1 NEXT in adjacent binder

[6] 10 x T1 NEXT in adjacent binder

[7] 3 dB margin

2.1.3.2 Crosstalk

Actual interference is due to the effect of other signals (ADSL or non-ADSL) on the data
pattern. With each increase in distance, the original signal level drops, making it more susceptible to *crosstalk* caused by these outside interferers. Crosstalk is the seepage of energy
from a signal along one copper pair to another and is the primary type of interference.

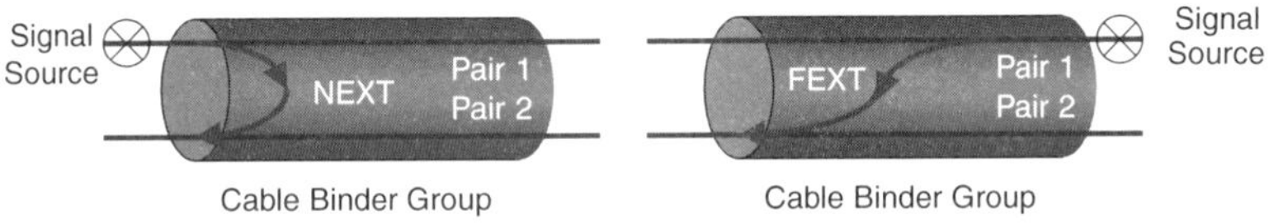

FIGURE 2.16.

Crosstalk

As depicted in Figure 2.16, there are two types of interference. The first, NEXT, is the effect
of another signal transmitted in the same direction as the original. The strength of this second signal (Pair 1) makes NEXT the more serious form of crosstalk since the received signal on the primary (Pair 2) is weak by this time. The second type, Far End Crosstalk (FEXT)
results from a signal transmitted in the opposite direction (Pair 1) from that of the original.
It is diminished by the time it reaches the primary's received signal (Pair 2).

ADSL is also subject to interference from external sources in the same frequency spectrum
(for instance, AM radio with a frequency overlap between 560 Khz and the ADSL maximum). DMT will adjust its bit density as described earlier or, in severe cases, cut off carriers
entirely. The solution for CAP is to adjust the carrier constellation pattern (that is, from
256-CAP to 64-CAP).

NOTE

DMT is able to change transmission speeds in increments as little as 64 Kbps by
adapting individual carriers, while CAP is capable of supporting a lesser number
of bandwidth increments.

In case of UADSL, the most interesting cause of interferences is due to phoneset events
such as ringing, on/off hook, and local noise generators. This is due to the particular characteristic of UADSL supporting integrated capabilities of supporting vocal band (0-4KHz) as
opposed as CAP or DMT requiring external splitters. Remember that UADSL does not rely
on the traditional splitter design of DMT or CAP. Instead, it relies on in-line, low-pass filters
installed at the phone jack and possibly high-pass filters at the ATU-R. Thus the UADSL
signal affects and and is affected by different types of phonesets, requiring country-by-
country and vendor-by-vendor testing. ISDN phonesets will present additional problems for

UADSL due to the technology's use of a higher frequency spectrum. This further cuts into the available UADSL bandwidth.

The expected performance listed in Tables 2.4 and 2.5 is important, but probably more critical is the actual reach that might be expected over different wiring gauges and in the presence of some common interferers. This is a more valuable analysis, since it will equip the service provider with the information necessary to guarantee a given bandwidth to different segments of the subscriber base (depending on how far they are from the CO).

Tables 2.6 and 2.7 depict the maximum achievable distances at a given bandwidth for CAP and DMT in the presence of both ISDN and HDSL impairments and with a cross-section of baud rates and constellation points. Note that there are many other possibilities. Although the distances may seem limited at the upper end of the bandwidth spectrum, in fact, most ADSL services are delivered at bandwidths much lower than those depicted.

TABLE 2.6. CAP EXPECTED REACH (DISTANCES IN KFT)

Payload Bandwidth (Kbps)	Baud Rate and Points	Expected Reach against NEXT with 6 dB Margin for BER $< 10^{-7}$			
		24 ISDN BRI		24 T1 HDSL 2B1Q	
		26 AWG	24 AWG	26 AWG	24 AWG
8192	1088/256	8.4	10.7	8.1	10.3
6144	1088/128	9.9	12.6	9.6	12.1
4096	1088/32	10.8	13.7	10.5	13.3
3200	680/64	11.7	14.9	11.1	14.2
2240	340/256	12.0	15.4	10.8	13.9
1280	340/32	13.8	17.8	12.6	16.1
960	340/16	14.5	18.7	13.3	17.0
640	340/8	15.2	19.6	14.0	17.9

TABLE 2.7. DMT EXPECTED REACH (DISTANCES IN KFT)

Payload Bandwidth (Kbps)	Expected Reach against NEXT with 6 dB Margin for BER $< 10^{-7}$				
	10 ADSL		24 ADSL	24 T1 HDSL 2B1Q	
	24 AWG	26 AWG	26 AWG	26 AWG	
6144	12.0	9.0	7.6	6.0	
	(CSA-8)	(CSA-6)	(CSA-4)	(Mid-CSA)	

Figure 2.17 graphically summarizes Tables 2.6 and 2.7.

FIGURE 2.17.

*Actual CAP and DMT
Performance*

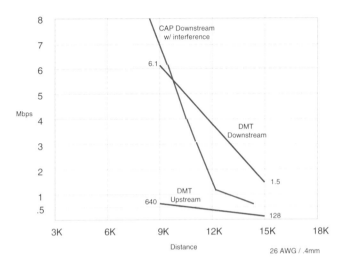

One area of debate, as yet unresolved, is which of the two encoding techniques is actually more likely to cause interference. DMT proponents make the point that subcarriers allow one to send out less power at a given frequency and contrast this to CAP, which requires more signal power at a given data rate and introduces additional interference. The opposite side of this argument is which of the technologies is more susceptible to interference. Here, CAP supporters take the upper hand and show how impulse noise will effect only a portion of the CAP spectrum but will effect multiple DMT subcarriers instead of a single one as expected (Figure 2.18). Shutting down carriers as described section 2.1.2.2 is a solution here. Since the vast majority of future installations will be based on DMT in any case, the discussion is all but moot.

FIGURE 2.18.

Effect of Impulse Noise

2.1.3.3 Deployment of CAP and DMT

Although this technology discussion is useful in understanding the different encoding options, the ultimate factor leading to success or failure will be how well the technology performs in actual deployments. This will determine the size of the user population served in terms of distances and bandwidth. Within the press and vendor community, the magic number for ADSL seems to be about 8 Mbps in the downstream direction and 1 Mbps upstream in terms of maximum bandwidth, with tariffed services at bandwidths from about 128 Kbps to 7 Mbps. Table 2.8 contrasts published values for CAP and DMT. Note the use of the word *published*. In actual implementations, service providers are finding that actual achieved performance varies on a pair-by-pair basis. Actual performance figures are somewhat difficult to come by, however, since they lend insight into a given vendor's chip implementation and are therefore proprietary. This is coupled with the reluctance of the ILECs and PTTs to publish loop performance test results.

TABLE 2.8.   DEPLOYABLE CAP AND DMT BANDWIDTHS

	DMT	CAP	G.lite
Downstream Max	6.1 Mbps[1] 8 Mbps (short dist)	7.168 Mbps[2]	1.536 Mbps[4]
Downstream Typ	1.5 Mbps	1.5 Mbps	512 Kbps
Downstream Min	384 Kbps	640 Kbps	64 Kbps
Upstream Max	224 Kbps 768 Kbps (short dist)	1.088 Mbps	512 Kbps
Upstream Typ	384 Kbps	256 Kbps	128 Kbps
Upstream Min	128 Kbps	90.6 Kbps	32 Kbps
Granularity	64 Kbps	136 Kbps Upstream 136, 340, 680, or 952 Kbps Downstream depending upon bandwidth	32 Kbps
Gauge (AWG)	24	24	24
Distance	12,000 ft	18,000 ft[3]	18,000 ft

[1] ADI AD20msp90 and 918 chipsets

[2] Globespan RDT-X0-01 chipset

[3] Globespan 14 Apr 98 announcement raises this to 26,000 ft

[4] UAWG, TG/98-10R5, Universal ADSL Framework Document, April 1998

A better insight into realized performance may be gained by looking at published tariffs and bandwidths. In the majority of the cases, the maximum downstream bandwidth is 1.5 Mbps, allowing service to the greatest number of users. A provider offering 6 Mbps would not find it beneficial to offer this higher bandwidth if it could be delivered only to a small percentage of the potential ADSL subscribers.

In terms of power consumption, CAP and DMT are about equal, though statements have been made that one technology is fundamentally more power hungry than the other. If anything, current Issue Two DMT chipsets do consume more power than CAP, though this is expected to improve in the future. A reason for this DMT's Peak-to Average Ratio (PAR) is defined in ANSI T1.413. In contrast to the DMT's requirement of 28 dB for no clipping, CAP requires 11 dB for 64-CAP and 14 dB for 256-CAP. A higher PAR results in increased analog front end complexity in terms of circuit design and power consumption. In addition, additional bits are required for the D/A and A/D converters, also resulting in additional power.

2.1.4 Interference within CAP and DMT

Equal or of even greater importance than ADSL performance is the effect of the technology on pre-existing services within a cable binder group. This factor is probably the most critical in a service provider's decision making and in fact drives many of the service decisions made by CLECs to go with alternate encodings and services (such as SDSL's 2B1Q).

Prior to the deployment of ADSL, services such as HDSL and ISDN requiring an extended frequency spectrum across the copper loop had no competition in terms of interference with the exception of bridge taps and the like. This all changes with ADSL, and the ability of multiple services to co-exist is known as *spectral compatibility*. Service providers planning to deploy ADSL technology have conducted a great deal of testing in this area, both CAP and DMT define the maximum tolerable interference, and the chip vendors provide good documentation. Table 2.9 lists the effect of ADSL (both CAP and DMT) as well as other services on the local loop. The first figure in the table, 24.6 kft, is the expected reach of POTS traffic in the presence of no interferers. In the same way, the expected reach of ISDN is 16.8 kft.

TABLE 2.9. EFFECT OF INTERFERERS ON REACH

Impairment	Reach @ 6dB margin
−140 dBm/Hz AWGN Only	24.6 kft
24 ISDN	16.8 kft
24 ISDN + 10 CAP T1 ADSL	14.4 kft
24 ISDN + 10 DMT	12.1 kft
24 HDSL	11.7 kft
24 HDSL + 10 CAP T1 ADSL	11.5 kft
24 HDSL + 10 DMT	11.1 kft
24 CAP T1 HDSL (784 Kbps)	11.3 kft
24 CAP E1 HDSL (1.168 Mbps)	12.0 kft

TABLE 2.9. EFFECT OF INTERFERERS ON REACH *(CONTINUED)*

Impairment	Reach @ 6dB margin
24 CAP E1 SDSL (2.048 Mbps)	12.1 kft
24 CAP T1 ADSL	21.4 kft
24 DMT	11.8 kft

2.1.5 ADSL and ISDN

With all the debate surrounding the support of ISDN in the presence of ADSL deployments, this topic is worth a discussion in itself. Although not a topic of major concern in the United States, in some countries in Europe and the Pacific Rim, compatibility of ADSL and the installed base of ISDN is a major issue. The first CAP and DMT hardware and POTS splitters supported analog phonesets but did not support ISDN with its greater frequency requirements. Toward the end of 1997, the ADSL vendors began to take this problem in hand, and by the second half of 1998 DMT solutions were available. Although initially vendor-specific, ETSI has now defined a standard for ISDN support which will be adopted by any DMT vendors seeking to support ISDN. CAP is following a different path, and though solutions have been proposed, the lack of CAP deployment in any ISDN-centric countries places actual implementation in question.

Reviewing ISDN frequency requirements once again, the 2B1Q encoding used in most countries requires 80 Khz of spectrum, in contrast to 4 Khz for POTS. The situation is even more extreme in Germany, as the 4B3T encoding in use extends to 120 Khz. The ideal solution would be to have a single scheme for both variants, though this would be at the expense of bandwidth. For DMT, solutions proposed include:

- Shifting the upstream and downstream spectrum upward at the expense of distance,
- Cutting out a portion at the expense of bandwidth, or
- Developing a more creative approach based on embedding the ISDN signal within the ADSL data.

The preferred solution within DMT is to extend the upstream frequency spectrum. Normally, DMT relies on 32 tones in the upstream direction, spaced at 4.3125 Khz (although 1–5 and 32 are unused). In ISDN mode, this is extended to 64 tones, deactivating 1–32 and 64. Here, the downstream data will occupy tones 64 to 255. Detailed filter design will be an issue, since both 2B1Q and 4B3T respond differently. The former occupies less spectrum but is more sensitive to group delay distortion from sharp filters, while the latter is less sensitive but occupies additional spectrum. To simplify matters, vendors are expected to release a single solution supporting both 4B3T and 2B1Q. Note that this does introduce a performance penalty for the 2B1Q users.

An important point is support within the same binder group for both analog and ISDN splitters. This may not be possible due to the frequency shifts described earlier. Even constraining the problem to ISDN, care must be taken to implement the same shift and splitter design. Within a binder group supporting ISDN, all ADSL modems must implement ISDN mode even if some subscribers do not require ISDN service.

2.2 ATM and ADSL

Above the copper media and the various ADSL encoding techniques lies ATM, the data, voice, and video encapsulation of choice for the vast majority of ADSL installations. This discussion follows a layered approach and first takes a step back by first introducing ATM as a technology. It then looks at other physical layer technologies supporting ATM, since ADSL over copper loops is only one of a number of possibilities. Next it looks at the ATM adaptation layer and finally at the various signaling, traffic management, addressing, and dynamic routing techniques. Note that this basic discussion of ATM is not intended to be exhaustive—the reader is referred to the bibliography contained in the appendix for additional background on what is a very broad subject.

2.2.1 Basic Concepts and Background

Asynchronous Transfer Mode (ATM) is a technology designed to preserve the quality of service (QoS) requirements of multiple traffic types carried over a single link or network. At the CPE, for example, voice, video, and data traffic may be combined for transport, with the bandwidth, loss, latency, and jitter requirements preserved. ATM accomplishes this by segmenting all traffic types into 53 byte entities known as *cells* associated with a different QoS.

Traditional voice traffic (non VoIP), for example, is very intolerant of delay and jitter across the network. This traffic will therefore be assigned a QoS guaranteeing proper delivery. In contrast, most data traffic is somewhat tolerant to changes in network performance and may therefore be carried with a less stringent QoS. When combining different traffic classes across a single link, the class requiring the more demanding QoS will take precedence. Consider an ATM CPE mixing voice and data traffic (Figure 2.19). The CPE segments both traffic types into ATM cells, but those belonging to the data traffic will be held in a queue if there is voice traffic to be sent. On the receiving end, the CPE reassembles the original voice or data frames.

2

FIGURE 2.19.

Data Types into ATM
Cells

ATM is traditionally a connection-oriented technology, establishing a circuit between the source and the destination. The connections may be under network control in the case of Permanent Virtual Circuits (PVCs) or initiated by the subscriber. These Switched Virtual Circuits (SVCs) rely on a signaling protocol described in Section 2.2.5. The reason for the qualifier traditionally is that MPLS, described in Section 2.2.5, (described later) changes this. Contrast this to connectionless IP datagram traffic, where the routing protocol routes the packets on-demand across the network. One of the challenges of the last few years has been to integrate properly the connection-oriented ATM Layer with the connectionless IP Layer.

2.2.1.1 Why ATM Was Chosen as ADSL Transport

There has been a great deal of debate regarding the overhead of ATM in comparison to other technologies. This is due to the ATM cell structure, consisting of a 5-byte cell header identifying the connection in question among other things, and a 48-byte cell payload. ATM overhead is in addition to any higher layer protocol overhead or that at the transmission layer. However, this loss in efficiency is more than made up by ATM's ability to meet the QoS needs of the different traffic types where required.

This last point is critical, since there will be many environments where ATM is unnecessary. For example, an ISP with a data-only service might be better served by implementing a pure packet over SONET/SDH backbone. Alternatively, a traditional service provider offering leased lines, Frame Relay, voice transport, and data will require ATM for its traffic management capabilities. More recently, some providers have begun to deploy multiservice offerings (voice and data integration) across ATM. The question, then, is why ATM was selected as the transport of choice for ADSL in the local loop, and why Frame Relay would not have sufficed.

In fact, the first encapsulation in use across ADSL was frame, and until the first half of 1997, the jury was out on which would dominate. The ISPs and some vendors favored frame due to its simpler implementation and lower overhead, while the service providers (ILECs/

PTTs) favored ATM due to its circuit-oriented nature and compatibility with recently deployed ATM access and core networks. Realistically, either technology would meet the needs of the majority of the services proposed, so the question was which of the two would gain market momentum (in the same way that the market decided in favor of DMT over CAP).

The direction was all but guaranteed when the major vendors all proposed ATM-based solutions into the major ILEC and PTT infrastructure bids, the most visible being the Joint Procurement Contract (JPC) in the United States. However, the use of ATM in the loop does introduce complexities in guaranteeing proper QoS mapping between it and the IP Layer,. In addition, the designers of ATM-based ADSL hardware sometimes did not have an appreciation for the requirements of the data traffic. This was just history repeating itself, mimicking deficiencies in the first generation of ATM WAN switches where buffering and congestion management were inadequate. In ATM's favor, the technology does support some traffic types above and beyond those capable with a frame-based encapsulation.

If one considers the ADSL local loop in the multiservice context, as opposed to a method of delivering only data, the advantages of ATM become clear. Just as ATM may be delivered over fiber, coaxial cable, or even wireless, it may be delivered via ADSL modems. Given proper CPE and support within the DSLAM, nothing precludes the delivery of CES/VTOA to ADSL subscribers, or the transport of MPEG2 video directly over the ATM Layer (Figure 2.20). A number of providers in fact plan these types of services, in addition to the more traditional Internet access. If ADSL's asymmetry is acceptable to the customer, the technology becomes a viable method of delivering integrated services to branch offices and to even the home.

FIGURE 2.20.

Multiservice from CPE

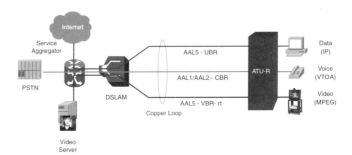

2.2.1.2 Alternatives to ATM

Although ATM is the technology of choice across the ADSL loop, alternatives exist once the traffic reaches the aggregation point. For example, multiple DSLAMs may be aggregated within a large CO and then backhauled over a packet over SONET/SDH backbone (Figure 2.21). Or, the subscriber PVCs/SVCs may enter the DSLAM, traverse an ATM access network to an aggregation point, and then undergo reassembly for transport across a packet backbone. This is a likely scenario, where an ISP or corporation attached to an ILEC's ADSL service forwards the subscriber across its internal packet-based backbone.

FIGURE 2.21.

Combining ATM and SONET/SDH

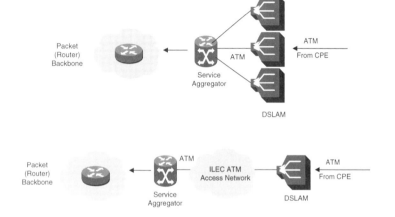

2.2.2 Physical Layer

The physical layer provides the actual connectivity for the higher layer ATM traffic, with options including fiber, copper, and wireless. It consists of the actual transmission systems along with the necessary glue to allow the ATM cells to be carried over the system in question. In order to accomplish this, it actually consists of two sublayers. The first is Physical Media Dependent (PMD), which provides for the actual line coding over the SONET/SDH, PDH, wireless, or ADSL system in question. Above this sublayer is Transmission Convergence (TC), responsible for cell scrambling, multiplexing, and HEC generation/verification. ADSL, operating at the PMD sublayer, and peering with any other transmission systems in use, is another option for ATM deployment.

Since ADSL is just one link within an end-to-end deployment, what are the other ATM physical layers you may expect to run across (realizing that ATM is one of a number of technologies deployed in the LAN and WAN)? In an environment where there is an ATM network on the subscriber side of the CPE, common link speeds and media types include:

- 622 Mbps (OC12/STM4) over single-mode fiber; 155 Mbps (OC3/STM1) over twisted pair; multimode and single-mode fiber; and, in a few cases, ATM25 (described below) or 2.4 Gbps (OC48/STM16).

- The SONET/SDH-based 2.4 Gbps, 622 Mbps, and 155 Mbps systems are very much like those deployed in the WAN. Other WAN interfaces include the PDH-based 45 Mbps (DS3), 34 Mbps (E3), 1.5 Mbps (DS1), and 2 Mbps (E1).

- A more recent interface is Inverse Multiplexing Over ATM (IMA), where multiple DS1/E1 links are combined into a single logical channel.

- The Frame-based UNI (FUNI) is less relevant in the ADSL space, since it is intended as an access link in the same way as ADSL.

The typical DSLAM will connect with an ATM switch via OC3/STM1, DS3, and in some cases, E3. If deployed in low-density environments or within a DLC/FSAN, DS1/E1 or the IMA is an option as well. A typical end-to-end ATM infrastructure could consist of ADSL in the local loop, SONET across the ATM access network, and PDH supporting the final link to an ISP. Table 2.10 lists the various physical media types.

TABLE 2.10. ATM PHYSICAL MEDIA OPTIONS

Media	Bandwidth	Encoding	Use
SM Fiber	155 Mbps (OC3/STM1)	SDH	Campus/ WAN
	622 Mbps (OC12/STM4)		
	2.4 Gbps (OC48/STM16)		
	(higher rates possible)		
MM Fiber	155 Mbps	SDH	Campus
	622 Mbps		
Coaxial	34 Mbps (E3)	PDH	WAN
	45 Mbps (DS3)		
	2 Mbps (E1) (or nxE1 via IMA)		
	1.5 Mbps (DS1) (or nxDS1 via IMA)		
Twisted Pair (UTP5)	155 Mbps	SDH	Campus/ Residence
	25 Mbps	4B5B	
Copper Pair	8 Mbps–32 Kbps (approx)	CAP/DMT	WAN
Wireless	155 Mbps–2 Mbps (approx)	Varies	Campus/WAN

Note: Not every defined ATM physical media is listed.

ATM25 deserves some discussion, since it is one of the service interfaces at the ATU-R (the other being Ethernet). Originally proposed for the campus environment, it never became a major factor even after concerted effort. This was due to the dominance of LAN switching, and where ATM was required to the desktop, most users opted for 155 Mbps. The interface was given a second lease on life when service providers and their ADSL equipment vendors decided that it would be a good demarcation interface between the ADSL service and the subscriber in those locations where the service provider controlled the CPE. The view was that it would be easier to manage and tariff than Ethernet, in addition to being able to support video delivered to STBs. However, even in those places where the ATU-R is provider-owned, ATM25 is in the minority. Nevertheless, it will succeed in this application space, and a number of telcos have in fact mandated ATU-Rs with both ATM25 and Ethernet interfaces, the former primarily intended for video.

2.2.3 ATM Layer

Above the Physical Layer is the ATM Layer, responsible for multiplexing the 53 byte cells over the physical media. As introduced above, these cells consists of a 5-byte header and a 48-byte payload. The header contains routing, flow control, and error control information, while the data payload in most cases contains the actual end-system data. Depending on whether the cell in question is generated at the User-Network Interface (UNI) or the Network-Network Interface (NNI), the header contents will change. The UNI is the ATM link between the subscriber and the first-hop ATM switch, while the NNI connects ATM switches in the network core. Figure 2.22 depicts the ATM cell header at the UNI; The NNI just adds additional VPI bits while eliminating the GFC field.

FIGURE 2.22.

ATM Cell Header

GFC: Generic Flow Control
VPI: Virtual Path Identifier
VCI: Virtual Channel Identifier
PTI: Payload Type Indicator
CLP: Cell Loss Priority
HEC: Header Error Check

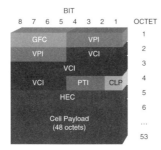

2.2.4 ATM Adaptation

The next function within ATM is data adaptation, probably the most interesting sublayer since it is here that the higher layer data types are adapted for transport within the ATM cells and where the actual Segmentation and Reassembly (SAR) is performed. There are five ATM adaptations defined, known as *ATM Adaptation Layers (AAL) 1–5*. AAL1, and in the future, AAL2, are usually considered the adaptations for constant bit rate (CBR) and voice traffic (though voice in fact is far from CBR). AAL5 is the data adaptation, while AAL3/4, designed for SMDS/CBDS, have all but disappeared and will therefore not be covered.

2.2.4.1 ATM Adaptation Layer 1 (AAL1)

AAL1 is designed for the transport of higher layer services requiring very tight constraints of data loss, latency, and jitter across the ATM network and is therefore well suited for circuit emulation. It is the adaptation associated with CBR traffic, though it is also used for real-time variable bit rate (VBR-rt) such as voice. AAL1 was also originally proposed for video (MPEG2) traffic, though AAL5 has taken this role. It accomplishes this by providing for timing recovery in the cell header and maintaining 64 Kbps channel structure in the payload. Methods of timing recovery include Adaptive Clocking and the Synchronous Residual Timestamp (SRTS). An ATM CPE will prioritize AAL1 traffic over all other adaptations, and ATM switches will feed this traffic through very small buffers, minimizing latency.

2.2.4.2 ATM Adaptation Layer 2 (AAL2)

AAL2 is a newer adaptation more suited for lower-speed connections and VBR sources. A problem with AAL1 manifested itself at lower data rates and low numbers of voice circuits. In this environment, the amount of time to segment a given voice circuit would be unacceptable. AAL2 solves this problem by defining *minicells*, which cut down on segmentation time. Though only recently defined, AAL2 is expected to find greater use in the future, especially over lower-speed ADSL connections.

2.2.4.3 ATM Adaptation Layer 5 (AAL5)

Data traffic is transported via AAL5, providing for no timing recovery. Service categories supported by AAL5 include:

- Unspecified bit rate (UBR) where the network provides minimal QoS guarantees. UBR with Early Packet or Partial Packet Discard (EPD/PPD) is sometimes referred to as UBR+.

- UBR-Weighted (UBRw), where the network handles traffic differently based on Diff-Serv (IP ToS) marking.

- Guaranteed Frame Rate (GFR), where the network makes a guarantee as to the minimum throughput.

- Available bit rate (ABR) where a feedback-based flow-control loop is implemented between the ATM CPE and the switch, or between ATM switches.

Unless a subscriber specifically requires other adaptations, such as AAL1 in support of VTOA, all traffic will be carried as AAL5. In addition, some DSLAMs support only a single cell buffer, precluding any traffic guarantees beyond those offered by UBR. AAL5 is also the most efficient adaptation in terms of data transport, since there is very little overhead in converging the traffic for segmentation. Figure 2.23 depicts the various layers involved when adapting user application data into AAL5, through the ATM SAR function, and onto the physical media. Note the increasing (albeit necessary) overhead with each layer.

FIGURE 2.23.

ATM Adaptation

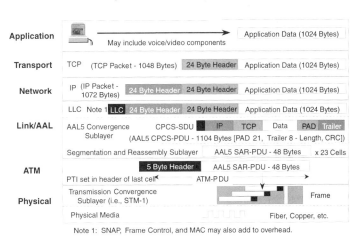

Note 1: SNAP, Frame Control, and MAC may also add to overhead.

2.2.5 Connection Types and Signaling

Within ATM, a single circuit connecting two points and supporting a single service category is known as a *virtual circuit (VC) or a virtual circuit connection (VCC)*. Multiple VCs may be grouped for management purposes into a *virtual path (VP) or virtual path connection (VPC)*. These VCs and VPs are identified via the VCI and VPI fields in the ATM cell header. Between an ATM CPE and the first-hop switch, every circuit is identified via a unique VCI/VPI combination. The same is true for the connections between the ATM switches, though the actual VCI/VPI values have only local significance—their values change when the circuits are cross-connected at the ATM switch. Some VCI values have significance, such as VCI=5, which is used for ATM signaling.

Since ATM is connection oriented as opposed to connectionless, some mechanism is required to inform the network to establish the virtual circuit from the source to the destination. There are two types of such mechanisms:

- A *permanent virtual circuit (PVC)* is one established via network management and is expected to be of long duration.
- In contrast, a *switched virtual circuit (SVC)* is one established under control of the CPE. This is an on-demand connection and may be short- or long-lived.

In the campus environment, the majority of ATM deployments rely on SVCs, while in the WAN, most networks support only PVCs at present. As will be described in Chapter 4, the majority of the ADSL service models require only PVCs, relying on higher layer protocols to establish connectivity dynamically from source to destination. Other types of connections exist.

One example is *Soft PVCs*, where the network manager preconfigures only the first and last hops, relying on the network to establish the connections dynamically. This minimizes the amount of provisioning required and allows the PVCs to follow more optimal paths than those that may be manually configured. *SVC tunneling* is a technique where the signaling channel (VCI=5) may tunnel across a non-SVC capable core via VPs. Soft-PVCs and SVCs rely on ATM routing, described in Section 2.2.8, to determine the best path between the source and the destination.

Probably the most radical alternation of the connection-oriented ATM model is in the form of Multiprotocol Label Switching (MPLS) as shown in Figure 2.24. Here, the ATM cell header is redefined to contain a label carrying Layer 3 destination information. MPLS-enabled ATM switches (known as Label Switch Routers) forward this Layer 3 information as opposed to setting up connections. The various steps taken in forwarding a packet across the backbone are outlined across the bottom of the figure. Since incoming data from multiple sources may need to be forwarded across a single outgoing interface, the MPLS-enabled switches must implement a feature known as *VC-Merge*, in effect operating in packet mode. Switches exchange Layer 3 reachability information via a Label Distribution Protocol, used to build the forwarding table in each. MPLS is relevant to ADSL in that it may form the basis of core connectivity. Even more relevant are MPLS VPNs, which are detailed in Chapter 4.

FIGURE **2.24.**

MPLS

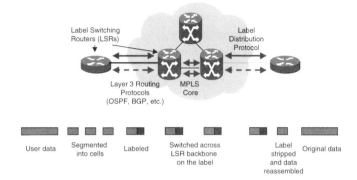

Signaling of SVCs relies on the ATM Forum's Signaling 4.0 specification. This describes the types and formats of the messages exchanged between the ATM-attached router or PC and the first-hop ATM switch. One of the most important pieces of information exchanged between these two systems is the type of connection in terms of QoS and bandwidth. This is covered in greater detail in Section 2.2.7. An earlier version of the standard, the UNI 3.1 (and the even earlier 3.0), is still implemented in some places. The subscriber may request different connection types depending on the application requirement:

- *Point-to-point (pt-pt)* connections are from a single source to a single destination.

- *Point-to-multipoint (pt-mpt)* allow a source to send data simultaneously to a group of destinations.

Under development are techniques to handle Multipoint-to-point (mpt-pt) where data converges on a single destination (much like VC-merge) and Multipoint-to-multipoint (mpt-mpt) forming a full mesh. At the ATM switch, the signaling message is handed off to the Private Network Network Internetworking (PNNI) protocol, described in the next section.

> The IETF is making an effort to define the set of capabilities required of data-centric CPE when establishing SVCs. These are described in RFC-2331.

2.2.5.1 VP and VC Switching

ATM switches may in fact implement different forms of switching. As Figure 2.25 depicts, a VP switch will act only on the virtual paths between sources and destinations (or other switches), switching via the VPI field. In contrast, a switch implementing VC switching will examine the Virtual Circuits within the VP, switching between Virtual Paths. Although almost every switch implements VC switching, management via the VPI and the use of VC switching are still important within service provider networks to optimize provisioning. In addition, the VPs play an important role in some of the ADSL switching models as introduced in the following sections.

FIGURE **2.25.**

VP and VC Switching

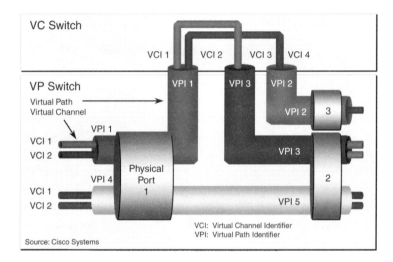

2.2.5.2 PVC Connectivity

PVCs are the most basic form of connectivity between source and destination. Here, none of the entites along the data path—the CPE, DSLAM, Aggregator, and any ATM switches—is required to implement ATM signaling (Figure 2.26).

FIGURE **2.26.**

PVC Connectivity

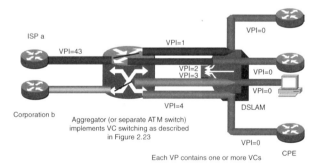

Network management handles provisioning of PVCs between the CPE and, while connections between switches may rely on PVCs (routed via the IISP or PNNI) or more commonly, the Soft-PVCs.

 NOTE

Most ADSL implementations today rely on PVCs for the most part, using Layer 3 techniques such as PPP for dynamic session establishment.

2.2.5.3 SVC Connectivity

In contrast to the permanent connections described in the previous section, SVCs allow connections to be established on-demand between source and destination or between the source and some intermediate point across the network, such as the Aggregator. They require a signaling entity in each network element capable of generating and responding to ATM signaling requests (Figure 2.27).

FIGURE 2.27.

SVC Connectivity

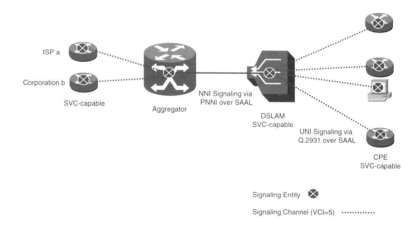

These messages are almost exclusively based on one of the ATM Forum's UNI specifications such as the UNI 3.1 or Signaling 4.0 (ATM Forum, 1996a). Signaling is carried between the user and the first hop ATM switch (the DSLAM in this example) by encapsulating the Q.2931-based signaling command in the Signaling AAL (SAAL). This is then carried between switches via the PNNI which determines the optimal path (as described in Section 2.2.5).

Two barriers to the widescale implementation of SVCs include the lack of wide support for SVCs on public ATM backbones, although this is changing, and the lack of SVC support on most DSLAMs, a condition likely to be rectified by mid-1999. Uses of SVCs include

video-on-demand services, some of which require SVCs and voice where it may be ineffi-
cient to allocate bandwidth ahead of time. One advantage of full SVC support across the
network is that it does not rely on any complex VPI mappings or control protocols as
described in the following two sections.

2.2.5.4 Virtual UNI

The next type of connectivity between the DSLAM and Aggregator (or ATM switch) is
known as the Virtual UNI. Here, each user is assigned a separate VP at the ingress of the
DSLAM, although all of these are mapped to zero as shown in Figure 2.28. Within the
DSLAM, these are cross-connected via a VP switching function to multiple VPIs, usually
based on the customer's identity.

FIGURE 2.28.

Virtual UNI

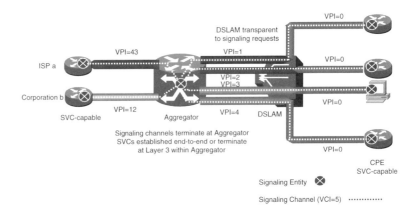

These VPs then terminate at the Aggregator. Within each VP, a signaling channel (VCI=5) is
transported from the CPE to the Aggregator, where it terminates, while user SVCs are established within the assigned VP. Thus the aggregator is required to terminate one signaling
channel for each subscriber, not a small task. Under this architecture, subscribers may initiate SVCs between the CPE and the Aggregator, even though the DSLAM has no concept
of SVCs.

NOTE

> Due to scalability concerns, this approach should be considered an interim until
> the wider deployment of SVC-enabled DSLAMs.

2.2.5.5 VB 5.2

Finally, an alternative approach has been proposed for the dynamic establishment of user connections. This is known as the *VB5.2* (Figure 2.29), a broadband version of the V5.2 protocol used between Access Nodes (AN) and the Local Exchange (LE) which contains the actual switching unit. It is therefore capable of managing multiple services including POTS and ISDN.

FIGURE **2.29.**

VB5.2

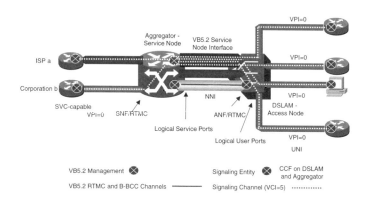

Under the VB5.2 architecture, all user signaling channels pass transparently through the DSLAM to the Aggregator, although these are no longer bound to specific VPs. Therefore, only one VP (or a small number) extends from the DLSAM to the Aggregator. The aggregator interprets the signaling requests and instructs the DSLAM via the RTMC channel to create the necessary cross-connect.

Although the VB5.2 interface has garnered some support within European service providers, it suffers the same scalability limitations as the Virtual UNI in terms of signaling channel terminations. It also introduces an additional management protocol between the Aggregator and the DSLAM which is not really needed. If most ADSL deployments were ATM-centric and focused on circuit switching (as with ISDN), the VB might prove useful. However, this is not the case, with the majority of ADSL very Internet focused. Here, Layer 3 techniques are more useful and scalable for control and session establishment.

2.2.6 Routing—The Network-Network Interface

Whereas the UNI handles connectivity and signaling between the subscriber and the first-hop ATM switch, a different protocol is required between ATM switches in the core of the network. This is the role of the *Network-Network Interface (NNI)*. A number of NNIs have been defined:

- Private NNI (PNNI)(ATM Forum, 1996b)—used in the corporate and service provider space.
- Interim Interswitch Signaling Protocol (IISP)—defined for use in the same solution space.
- Broadband Inter-Carrier Interface (B-ICI)—developed for connecting ATM providers.
- ATM Internetworking Interface (AINI)—instituted for interconnecting both private and public networks.

This book focuses on the PNNI, shown in Figure 2.30, as the best example of an ATM dynamic routing protocol.

FIGURE 2.30.

ATM Routing via the PNNI

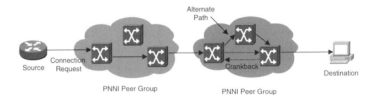

As with any routing protocol, the PNNI relies on the exchange of reachability information between ATM switches to construct a least-cost routing table. It is a link-state protocol, much like OSPF or IS-IS, but it also includes QoS awareness (possibly due to the connection-oriented nature of ATM) and automatic summarization into a hierarchy. Groups of switches under a single administrative domain exchange this information through PNNI Link State Advertisements (PLSA) and select a group leader. This leader injects a summary of reachability information into the next higher layer of the hierarchy. In this way, up to 104 layers of hierarchy are automatically created, allowing for massive scalability.

When a subscriber initiates a signaling request across the UNI, the first-hop ATM switch hands off the parameters (QoS, security, destination) into the PNNI. This protocol, with knowledge of reachability, forwards the request on the best known path across the network which will meet the signaled parameters. This process is repeated switch-by-switch across the network until the request reaches the destination. However, if the request reaches a dead-end in terms of a specific path being able to support the request, the protocol implements a technique known as *crankback* which will allow the request to attempt a new path. Since PNNI is QoS aware, the link state advertisements include details on the QoS capabilities of the links across the network.

2.2.7 ATM Addressing Plans

The creation of SVCs and the routing of calls via the PNNI is possible only if you have a way to identify devices connected to the ATM network. This is the role of an *ATM addressing plan*, where a unique identifier is assigned to every system connected to the ATM network. There are in fact two addressing plans in use. The more common *ATM End-System Address (AESA)* is used within both private and public networks and forms the basis for operation of the PNNI. A number of AESA formats have been defined depending on who is responsible for the addressing hierarchy (Figure 2.31).

FIGURE 2.31.

AESA Formats

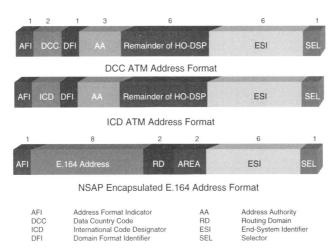

AFI   Address Format Indicator   AA   Address Authority
DCC   Data Country Code   RD   Routing Domain
ICD   International Code Designator   ESI   End-System Identifier
DFI   Domain Format Identifier   SEL   Selector

The second form of addressing, also shown in Figure 3.20, is based on the ITU administered E.164 address format and is equivalent to that used within ISDN numbering. This format is used only within public networks and relies on the B-ICI for routing if SVCs are implemented.

At the edge of public networks, some switch vendors support interworking between AESA and E.164 format addressing. For example, a corporation may be assigned a single E.164 address at the public network boundary, while using AESAs internally. If this corporation has multiple sites connected to the public network, destination AESAs may in fact be carried as subaddresses when a call is established across the public cloud. This provides a user internal to one site with the ability of signaling a connection of a user internal to a second site with the call traversing the public network.

Looking more closely at AESAs, the traditional address format is known as a *unicast* address, where a single source establishes a connection to a single destination. A second type of address format allows a caller to locate the nearest instances of services across the network. This is known as *anycasting* and uses the group addressing format first defined in Signaling 4.0.

> Multicasting in the realm of IP multicasting is supported by opening multiple leaves of a pt-mpt VCC. There is no multicast-format ATM address as of yet, in contrast to IP's Class D addressing.

Efficient routing across a private or public network is made possible only by proper selection of an addressing plan. Over the years, there has been a lot of effort placed in this area, taking into account administrative domains, growth, and the type of summarization required by the PNNI. In the ADSL space, proper up-front planning of the addressing plan in use is critical. Chapter 5 outlines one example.

2.2.8 Traffic Management

Probably the most complex area of ATM as a technology is *traffic management*, the job of ensuring QoS for the voice, video, and data traffic carried across the network. Why the need for traffic management at all? If the network was never congested and subscriber applications could request as much bandwidth as required with no impact on other users, there would be no need. However, in any real network, this is far from the case. Trunks are oversubscribed, buffers within switches are not infinite, and the QoS requirements of the different application types must be preserved even on low bandwidth connections.

Proper traffic management (or lack thereof) at the ATM layer also has a dramatic effect on data traffic performance. TCP, the most common transport protocol (Layer 4) expects the network to react to congestion in a certain way as to preserve "goodput," the amount of good data which actually arrives at the destination, the internetworking layer. This requires proper discard mechanisms within the ATM switch and adequate buffering in both the switches and the edge devices.

2.2.8.1 ATM Service Categories

ATM defines a number of service categories related to ATM adaptations. The most stringent category is CBR, commonly associated with circuit emulation or voice traffic and also associated with AAL1. Less demanding is VBR, divided into Real-Time (VBR-rt) and non-Real-Time (VBR-nrt) depending on the application. VBR is especially suitable for voice and video traffic due to their burstiness and is mapped to either AAL1, AAL5, or the newer AAL2. Finally, internetworking traffic, including IP-based voice and video, may be mapped into either UBR or ABR depending on the QoS demands of the application.

NOTE

Of note is that campus ATM networks almost exclusively rely on UBR, while ABR finds use at the WAN boundary. This is due to the fact that a properly designed campus network should support application QoS without resorting to ABR.

Both the ATM Forum and the ITU-T have been active in defining ATM traffic management. Figure 2.32 describes these different service categories.

FIGURE 2.32.

ATM Service Categories

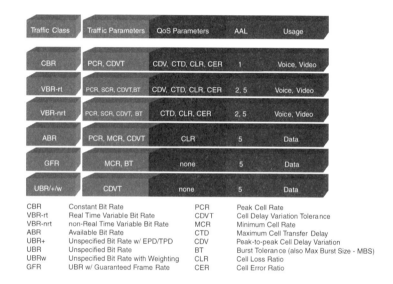

Traffic Class	Traffic Parameters	QoS Parameters	AAL	Usage
CBR	PCR, CDVT	CDV, CTD, CLR, CER	1	Voice, Video
VBR-rt	PCR, SCR, CDVT,BT	CDV, CTD, CLR, CER	2, 5	Voice, Video
VBR-nrt	PCR, SCR, CDVT, BT	CTD, CLR, CER	2, 5	Voice, Video
ABR	PCR, MCR, CDVT	CLR	5	Data
GFR	MCR, BT	none	5	Data
UBR/+/w	CDVT	none	5	Data

CBR	Constant Bit Rate	PCR	Peak Cell Rate
VBR-rt	Real Time Variable Bit Rate	CDVT	Cell Delay Variation Tolerance
VBR-nrt	non-Real Time Variable Bit Rate	MCR	Minimum Cell Rate
ABR	Available Bit Rate	CTD	Maximum Cell Transfer Delay
UBR+	Unspecified Bit Rate w/ EPD/TPD	CDV	Peak-to-peak Cell Delay Variation
UBR	Unspecified Bit Rate	BT	Burst Tolerance (also Max Burst Size - MBS)
UBRw	Unspecified Bit Rate with Weighting	CLR	Cell Loss Ratio
GFR	UBR w/ Guaranteed Frame Rate	CER	Cell Error Ratio

The goal of any traffic management system is to preserve the QoS requirements of traffic conforming to these service categories. There are a number of ways to accomplish this:

- Short queues and proper traffic prioritization along the end-to-end path help ensure that the needs of CBR and VBR-rt traffic are met.
- UBR, where few guarantees are made, relies on Early or Partial Packet Discard within the ATM switches (and is commonly referred to as UBR+ when these capabilities are present). Here, in the presence of congestion, a switch forced to discard one cell belonging to a higher-layer packet will discard all succeeding cells. This minimizes the number of wasted cells sent across the backbone and helps preserve the "goodput" introduced earlier. UBR-Weighted (UBRw) goes a step beyond this by treating traffic differently in terms of discard probability based on DiffServ (IP ToS) marking. A newer form of UBR known as GFR, for Guaranteed Frame Rate, attempts to make some guarantees as to QoS by specifying a minimum transmission rate.

- ABR in contrast relies on a sophisticated feedback loop between the subscriber, the ATM backbone, and the destination with a net result of guaranteeing bandwidth across the network. It operates proactively, informing the source of impending congestion. Under this rate-based mechanism, the source is now expected to reduce its transmission rate, with a goal of protecting the backbone. With all sources cooperating, each gets fair access to network resources.

> Proper operation of ABR relies on interfaces capable of buffering traffic equal to the amount of data in transit over the serviced link. If this was not the case, the switch would be incapable of buffering the traffic in times of link congestion. . . The amount of traffic in bits over the link, and therefore the amount of buffering required, are referred to as the Bandwidth-Delay (bw*rtt) product. They also play a role in proper TCP operation, described in Section 2.3. ABR is defined in the ATM Forum's Traffic Management 4.0 specification (ATM Forum, 1996).

Of importance is the interrelationship between the IP and the ATM Layers and its effect on traffic management. Until recently, these two layers had no coupling, and the only way to ensure proper operation of the internetworking layer was to perform traffic management at the ATM Layer. This included ensuring adequate buffering, implementing EPD/TPD for UBR+, GFR, or ABR, and watching oversubscription ratios across ATM trunks. At the TCP Layer, the use of large window sizes helped to preserve throughput across high-bandwidth long-delay links. What was really needed, however, was interworking between the IP and ATM Layers, allowing applications to request service categories other than UBR or ABR when required. Alternatively, given a single ATM VCC, proper queuing at the subscriber could help preserve application QoS requirements.

In today's deployment of IP applications it is possible to classify different application traffics based on their priorities using IP Type Of Service (ToS) bits: This can be done by the application itself or by a router with classification functions. Once classified, intermediate routers can perform routing decisions based not only on the canonical destination IP address but also on the specific application (i.e., TCP or UDP port) and the value of IP TOS, partitioning traffics based on priorities. This capability may be applied to ATM in one of two ways. The first relies on multiple VCCs, where VoIP or networked video, properly prioritized, may be carried over a VBR-rt VCC, while bulk data is mapped to a UBR+ or GFR virtual circuit. Alternatively, if a single VCC is used, it is still possible to preserve traffic priorities by implementing proper queuing at the CPE.

These techniques directly relate to implementations across ADSL. Given the need to support these different traffic types, an ADSL CPE could implement the necessary queuing and then map the applications into one or more VCCs. This of course requires support on the DSLAM. If the DSLAM has only a single set of buffers and if the ATM uplink is oversubscribed as expected, it will have no way to differentiate the time-critical traffic from the bulk data. Even if multiple PVCs were established from a subscriber, all traffic would flow into a single set of buffers. Therefore, the DSLAMs require at least two types of buffering: real-time and non-real-time. The uplink then connects to an ATM switch where we expect full traffic management.

2.2.8.2 Voice over IP (VoIP)

Of the possible applications to be offered across ADSL, one receiving a great deal of coverage is Voice over IP (VoIP). Here, the voice traffic is digitized and then encapsulated into IP at the ADSL CPE, or possibly at the PC. At the CO, it is either gatewayed into the PSTN or further transported across the IP backbone. However, support for VoIP across ADSL does introduce some concerns regarding QoS support at the DSLAM and within the provider's network. Even if the CPE is capable of prioritizing traffic at the IP Layer, interleaving the voice traffic into the data stream as to reduce delays, the DSLAM has no concept of this prioritization. It operates at the ATM layer, and if equipped with only a set of best effort (UBR) buffers, real-time (VoIP) and non-real-time (all other) traffic from all subscribers is interleaved and placed on the single upstream ATM trunk. If the ADSL provider is careful not to oversubscribe users, this may not be a problem, but this is rarely the case. As with other data services, one element on the path to profitability is oversubscription, and in this environment QoS for any given subscriber cannot be guaranteed.

Even if the statement that each and every subscriber *should* receive a fair share of the available bandwidth could be made, there is no way to guarantee that the VoIP traffic will not be delayed by a large quantity of non-real-time traffic. This is not to imply that VoIP will not operate at all, just as VoIP many times delivers acceptable quality over dial-in Internet access. It does imply, however, that the service provider cannot offer any quality guarantee, important for wider acceptance and especially for corporate users.

The way around this is to implement differentiated buffering in the DSLAM. For example, some DSLAMs support both UBR and VBR-rt (or CBR) buffers. The question then is how to map the different forms of IP traffic at the CPE into PVCs with different QoSs. Sufficiently intelligent CPE now exists with this capability, mapping the VoIP traffic onto a CBR or VBR-rt PVC, for example, while mapping all other traffic to UBR (or even ABR if available).

A second, less elegant solution, would be to map sets of subscribers to different DSLAMs, each with its own uplinks. For example, casual users with no VoIP guarantees could be served from a DSLAM with high oversubscription, while those with more stringent requirements, such as VoIP or just bandwidth guarantees could be served from a separate system.

2.3 Frame—An Alternative

Although the vast majority of ADSL deployments are based on the ATM encapsulation, there are some DSLAMs operating in packet mode. In addition, the ADSL Forum has released an architecture document for packet mode operation (ADSL Forum, 1997a). Note that the discussion here relates to packet across the ADSL loop. Nothing precludes an ATM-based DSLAM vendor from implementing a SAR in the DSLAM uplink, feeding FE or Frame into the access network or PoP. Alternatively, the core network upstream from the frame-mode DSLAM may be based on any technology. If an end-to-end architecture supporting L2TP tunneling is deployed as described in ADSL Forum, 1998b, the network must of course support this encapsulation.

Frame operation relies on HDLC or FUNI encapsulation across the ADSL loop, supporting any of the higher-layer protocols such as PPP, routing, or bridging. The FUNI (ATM Forum, 1997b) is no different than that defined for access to ATM over non-ADSL links. The header (see FUNI Encapsulation in Figure 2.33) encodes the VPI/VCI fields along with fields for congestion notification, cell loss priority, and a CRC. In the case of ADSL, the default combination is VPI=1, VCI=32, although other values may be provisioned. A vendor specific channel for management channel, if implemented, will be carried in VCI=33, while OAM cells are carried in frames with VPI=0 (ADSL Forum, 1998c). The data frame size is set to at least 1,600 bytes to allow for all types of Ethernet encapsulations. If supporting bridged traffic, the FUNI will encapsulate it IAW RFC-1483's LLC format. Alternatively, VC-mux may be used to support other traffic types such as routed data. PPP traffic must be carried as VC-mux format. HDLC is an older encapsulation, defined in Simpson (1994). Within ADSL, it must operate in bit-synchronous mode, with the header as described in HDLC Encapsulation in Figure 2.33.

Figure 2.33.

Frame Encapsulation

Frame Encapsulation over ADSL

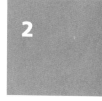

2

(a) **FUNI Encapsulation**

(b) **HDLC Encapsulation**

Frame mode does have some advantages in that the complex QoS interworking between Layer 3 and the ATM Layer is eliminated. In addition, there is no ATM cell overhead. These advantages, however, are more than balanced by provisioning problems. A frame-mode DSLAM must have some method of mapping users to an uplink, via either 802.1q or some proprietary mechanism emulating VLANs. Scaling this into the thousands of subscribers while guaranteeing security presents its problems. Between users, QoS is much more difficult to enforce given current Layer 2 tools. This mode also precludes the delivery of native ATM services such as VoIP and MPEG2 video.

Probably the most critical issue, however, is the desire of ADSL providers to use their ATM access networks for service delivery. Thus some conversion from frame to ATM would have to occur somewhere in the access network. Where this should occur in terms of manageability and economics is sometimes a matter of debate, but the reality is that the carriers have deployed ATM mode, with the SAR at the subscriber. Chapter 4 describes frame over ADSL architectures in greater detail.

2.4 Data Encapsulations

ATM signaling, routing, addressing, and traffic management are only a support framework for higher-layer services. These services may include voice, video, and data transport. In the latter case, a number of methods of transporting data across the ATM network have been defined, some optimized for the campus and others for the WAN. Those optimized for the campus include LANE and MPOA, while PPP over ATM and MPLS are found in the WAN. RFC-1483 bridging and Classical IP based on RFC-1577 see use in both the campus and the

WAN.This section briefly outlines the data models deployed and expected to be encountered as part of an ADSL service.

One of the challenges has been to support higher-layer application requirements across the ATM backbone. Only recently has there been any integration between Layer 3 QoS (IP Precedence, RSVP, and the like) and the ATM service categories. Although these capabilities will evolve in the future, different data models allow this interworking to a lesser or greater extent. They should therefore be considered in the context of how well they support these requirements in the LAN and WAN, especially as IP-based real-time applications become more commonplace.

2.4.1 Bridging

Bridging is the simplest method of interconnecting two or more internetworking devices and relies on Link Layer (Layer 2) addressing for communications. In the case of Ethernet, this is the 6 octet MAC address familiar to those who have ever worked with Ethernet NICs. Although large networks may be based on bridging, routing (described in Section 2.5) is more scalable.

Translated to the ATM domain, RFC 1483 (Heinanen, 1993) outlines a method by which multiprotocol traffic may be bridged across the ATM (and therefore ADSL) local loop between a host and a LAN switch or router. Since it provides for no address resolution mechanisms, unlike LANE (described in section 2.4.3), it relies on PVCs. In the WAN or MAN, where SVC support is less common, this is not a major problem, and RFC 1483 bridging has been deployed in a number of instances in support of metropolitan LAN extension services. It is less common in the campus, where SVCs are desirable and LANE, MPOA, or the Classical Model better meet requirements.

Figure 2.34 depicts three ATM-attached devices—a router, a PC, and a LAN switch—connected via ATM PVCs to a centralized router. Here, a centralized router terminating the ATM VCCs, may forward data at the bridging layer or, alternatively, may route traffic between connected subnets. If a bridge instead of a router were deployed, it would be capable only of bridging layer forwarding. Bridging across ATM is always on, with no provisions for user authentication or start/stop accounting unless this takes place at the ATM layer.

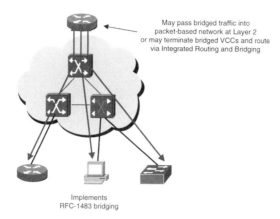

FIGURE 2.34.

Bridging over ATM

May pass bridged traffic into
packet-based network at Layer 2
or may terminate bridged VCCs and route
via Integrated Routing and Bridging

Implements
RFC-1483 bridging

2.4.2 Classical IP

The first routed model deployed, and still quite viable, is Classical IP and ARP over ATM (Laubach, 1998). This is a Layer 3 model, where the ATM core is surrounded with routers. Although the title might lead you to believe otherwise, the architecture in fact supports multiple Layer 3 internetworking protocols such as IPX and AppleTalk. When used in this context, it is sometimes referred to as the *Classical Model* as opposed to Classical IP. The technique also supports IPmc via proper router software of the deployment of multicast servers. Two approaches are documented as part of the Multicast Address Resolution Service architecture (RFC 2022) or a simpler router-based approach based on PIM Sparse Mode (RFC 2362).

Given proper QoS capabilities on the router ATM interfaces, Classical IP also supports QoS interworking since the router-ATM interface has visibility into ATM QoS. Its Classical IP is deployed in both the LAN and WAN, though its use in the LAN is on the decline with the deployment of LANE/MPOA. Though not the primary data model across ADSL, it will see some use.

The actual data encapsulation technique as defined in RFC 1483, on which Classical IP relies, however, forms the basis for bridging across ADSL. Figure 2.35 depicts a typical Classical IP implementation.

FIGURE **2.35.**

Classical IP

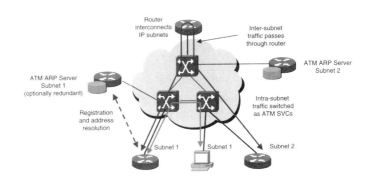

A host wishing to reach a given IP destination in the same IP subnet (Subnet 1) will first query the ATMARP server for the destination's ATM NSAP address. It will then proceed to establish an SVC to the destination. Alternatively, a host wishing to reach a destination in a different subnet (Subnet 2) will forward its traffic to the default router over an SVC or PVC. This two hop forwarding between subnets is one of the problems identified with Classical IP, leading to a requirement for either MPOA or MPLS.

2.4.3 LANE and MPOA

In the campus, and sometimes in the MAN, *LAN Emulation (LANE)* (ATM Forum, 1997f), and more recently, *Multiprotocol over ATM (MPOA)* (ATM Forum, 1997e) are encountered. LANE is a bridged model, attempting to emulate the characteristics of a LAN segment across the ATM backbone. It does this rather successfully through the deployment of a complex control protocol between the edge devices, known as *LAN Emulation Clients (LECs)*, and the control entities, known as *LAN Emulation Servers (LESs)*.

Broadcast and Unknown Servers (BUSs) emulate the broadcast/multicast capability of a LAN segment. Each Layer 2 domain, commonly associated with an IP subnet (to use IP as an example, though LANE is inherently multiprotocol capable), is termed a *Virtual LAN (VLAN)*. Routers interconnect the VLANs.

MPOA builds upon LANE by allowing the edge devices (commonly LAN switches) to establish cut-through connections from one VLAN to another. This avoids the congestion commonly associated with centrally deployed routers. As with LANE, it relies on a control protocol between MPOA clients and servers. In an ADSL deployment, the LANE or MPOA domain will terminate at the ADSL CPE.

Figure 2.36 depicts a typical LANE environment, with the various LECSs, LECs, LESs, and BUSs. A host (LES) in Subnet 1 wishing to forward traffic to another host in the same subnet first queries the LES for the destination's ATM NSAP address (unless this information is previously cached). It will then proceed to establish an SVC to the destination. However, if the destination is in a different subnet (Subnet 2), it will forward all traffic to the default router (itself an LEC) for forwarding to the destination.

FIGURE 2.36.

LAN Emulation

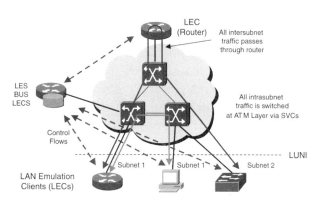

MPOA, as depicted in Figure 2.37, operates in much the same way with the exception that the address resolution mechanism via the MPS allows hosts (MPCs) to establish direct SVCs between different subnets. MPCs must therefore be equipped with ATM interfaces with Layer 3 intelligence. MPOA also provides for a default forwarding path between subnets if a direct SVC is undesirable (that is, low traffic volume or policy issues).

FIGURE 2.37.

Multiprotocol over ATM

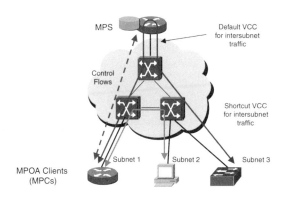

2.4.4 Point-to-Point Protocol (PPP) over ATM

Point-to-Point Protocol (PPP) over ATM (Gross, 1998) is a more recent encapsulation, primarily implemented as part of ADSL. This technique relies on RFC 1483, operating in either LLC-SNAP or VC-Mux mode. A PPP-compatible CPE will encapsulate the PPP session based on this RFC for transport across the ADSL loop and DSLAM.

As depicted in Figure 2.38, PPP operates in a true point-to-point mode, where all VCs terminate at a router, and traffic is routed between users. Initially, PPP over ATM will rely on PVCs, since the actual higher layer session establishment is still dynamic and based on authentication between the user and an external RADIUS or TACACS+ server.

FIGURE **2.38.**

PPP over ATM

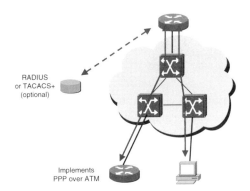

RADIUS
or TACACS+
(optional)

Implements
PPP over ATM

In the future, this encapsulation could rely on SVCs as well, given adequate support in the service provider's core. Although PVCs are adequate for PPP operation, a desire to minimize preprovisioned resources may lead some providers to opt for SVCs. PPP over ATM is described in greater detail within Chapter 4 since it will be the primary data encapsulation across the ADSL loop.

A newer architecture is based on Multiprotocol Label Switching (MPLS) (Callon, 1997). MPLS redefines the way in which ATM operates, and instead of relying on signaled connections between the source and destination, operates in a connectionless mode. Here, the ATM cell header is redefined to contain a label carrying Layer 3 destination information.

MPLS-enabled ATM switches—known as *Label Switch Routers (LSR/)*—forward on this Layer 3 information as opposed to setting up connections. Since incoming data from multiple sources may need to be forwarded across a single outgoing interface, the MPLS enabled switches must implement a feature known as VC-Merge, in effect operating in packet mode. Switches exchange Layer 3 reachability information via a *Label/Tag Distribution Protocol (LDP/TDP)*, used to build the forwarding table in each device.

MPLS is relevant to ADSL in that it may form the basis of core connectivity. Even more relevant are MPLS VPNs, detailed in Chapter 4. Looking back to Figure 2.24, a LDP is used to build a table within each LSR. This table corresponds to Layer 3 destinations and is used to label user data entering the LSR backbone. The LSRs therefore switch on this label instead of the full IP header, a great improvement in efficiency. At the egress point of the LSR backbone, this label is removed.

2.4.5 Native ATM Services

Although the focus here may seem to be the support of internetworking traffic across ADSL, the success of the technology, at least in some locations, will also depend upon the deployment of *native ATM applications*, that is, applications that make direct use of ATM services while relying on an internetworking layer such as IP. This is especially true in areas where ADSL is considered a means to deliver multiple service types, replacing leased lines. Two major classes of applications are relevant:

- **The encoding of high-quality video across ATM**—In most cases, this will be MPEG-2, relying on AAL5 and a VBR-rt PVC or SVC, though AAL1 may be deployed as well. ADSL-based video services are initially planned for the residential environment, though they could be used within education and government as well.

  The ATM Forum defines the method by which the MPEG-2 Transport Stream consisting of multiple channels and a Program Clock Reference is encoded within AAL5, while the ITU-T defines the same for AAL1. Where video over ATM is deployed to the CPE, it would be carried in a separate PVC or SVC from the internetworking data traffic.

- **Circuit replacement, including voice transport**—ATM is especially suited for leased line replacement due to the guarantees CBR (and VBR-rt) place on QoS, and given proper support on the CPE and DSLAMs, ADSL may be used to deliver some services traditionally supported via fiber. In the case of ADSL, voice is probably the major driver.

  The ATM Forum has defined a complete voice architecture known as *Voice and Telephony over ATM (VTOA)* (ATM Forum, 1997). It includes provisions for dynamically establishing sessions, transport of signaling protocols between PABXs, and compression. VTOA relies initially on AAL1, but in the future, AAL2 should see greater deployment. Where VTOA is deployed, one could envision an ADSL CPE supporting both internetworking as well as native ATM applications. A phoneset or PABX will connect to an interface, feeding directly into the ATM/ADSL interface and bypassing Layer 3 as depicted in Figure 2.39. At this interface, both the voice and data traffic will be mapped into discrete PVCs (or SVCs) for transport.

FIGURE 2.39.

Voice and Telephony
over ATM

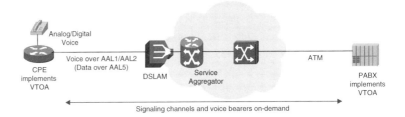

The DSLAM would of course have to provide differentiated buffering for the different VCCs to ensure QoS, and if connections are to be established on-demand, it must support SVCs. The voice traffic then flows across the ATM backbone to a VTOA-capable destination such as another router or ATM-connected PABX. One additional consideration is that video and voice applications rely on proper network timing support. This must be planned for in the design.

The video and voice applications described are initially deployed on what may be considered native ATM devices such as multiservice CPE or STBs. With the release of Windows 98 and an ATM API, the PC is now *ATM-enabled*. This will allow it to support ATM applications in parallel with traditional internetworking-based applications. In Chapter 5, native ATM video and voice configurations are both detailed.

2.5 The Network and Transport Layers

Having covered the ADSL physical and ATM layers, along with the various data encapsulations which ride above these layers, the next area of interest is the Network and Transport Layers, with an emphasis on the Internet Protocol (IP) due to its ubiquity in the Internet and within the majority of ADSL deployments. All data applications rely on network and transport layer protocols, along with the proper interworking between these layers and the lower layers.

NOTE   Other types of applications will exist across ADSL. In Section 2.2.9, the direct encapsulation of voice and video across ATM was described. Although these will see some use, they will remain niche applications in terms of subscribers served.

This section first introduces the IP protocol, which leads into a discussion of addressing, routing, multicasting, QoS, and security. Other protocols and techniques closely related to IP include DHCP, the DNS, and NAT. A next-generation IP, nearing deployment, is detailed as well, along with currently deployed tunneling mechanisms such as L2TP and L2F. One layer up from IP in the protocol stack is the transport layer. Here, the focus is on both the Transmission Control Protocol (TCP) and the User Datagram Protocol (UDP), describing where and how each is deployed.

2.5.1 The Internet Protocol

The Internet Protocol (IP) (Postel, 1981) lives at Layer 3 of the internetworking stack, above the Link Layer (in this case ATM) and below the Transport Layer. The IP header contains the necessary addressing information to allow a data packet to be routed across the Internet.

The concept of a network connecting computers was first described in the 1960s, and by 1969 the first nodes (switches) and hosts (computers) on the ARPANET were connected. Only later did the U.S. military discover that the technology used could form the basis of a survivable network for wartime. The first Layer 3 protocols in use were 1822 and Host-to-Host, while the Network Control Protocol (NCP) provided for transport service. It was not until the beginning of the 1980s that the Internet Protocol (v4) (RFC-791) as we know it today was first deployed. Initially, computer vendors were slow to adopt IP, but as the ARPANET (and its military outgrowth, MILNET) took hold, deployment became more widespread. A major hurdle in the form of an alternative protocol environment, based on the OSI model and promoted by the U.S. government as GOSIP, never took hold. As the early networks grew in extent, the term "Internet" came into wide use. Toward the end of the 1980s, with the completion of the global routing hierarchy, exponential user growth via dial-in, and the ensuing popularization of the Internet in the press, the dominance of IP was all but ensured.

2

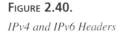

FIGURE 2.40.

IPv4 and IPv6 Headers

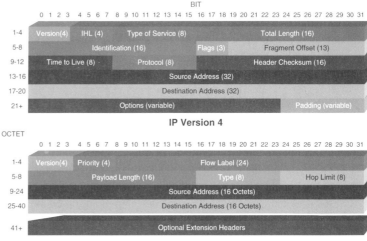

The top half of Figure 2.40 depicts the various fields within the IPv4 header. Their functions are as follows:

Field	Length	Function
Version	4 bits	IP version number—set to "4" for IPv4
IHL (Internet Header Length) words	4 bits	Length in 32 bit
Type of Service	8 bits	Parameters of QoS desired. Bits 0–2 define IP precedence ranging from 111 (network control) to 000 (routine). Bits 3–5 request bounds on delay, throughput, and reliability
Total Length	16 bits	Length of datagram in octets; max=65,635
		Host must accept min=576
Identification	16 bits	Used to identify fragments
Flags	3 bits	Bit 1—may or may not fragment
		Bit 2—last fragment or more fragments
Fragment Offset	13 bits	Tells where in datagram fragment belongs in units of 8 octets
Time to Live	8 bits	Max=255; decremented at each router
		If "0", then datagram is discarded
Protocol	8 bits	Layer 4 protocol (i.e., TCP)
Header Checksum	16 bits	Header only; recalculated at each router
Source Address	32 bits	IP source

Field	Length	Function
Destination Address	32 bits	IP destination
Options (optional)	Variable	Security, source routing, etc.
		Multiple options permitted
		Acted upon at each hop, the discussion of IP would not be complete without a look ahead to *IPv6* (Deering, 1995), sometimes known as *IP Next Generation*. IPv6 is the outcome of concerted efforts over the last 5 years or so to develop a next-generation IP protocol solving many of the deficiencies identified with the currently deployed IPv4. These primarily focus on security, auto-configuration, QoS, extended address spaces, header complexity (looking back to the various fields in the IPv4 header), and better processing of options. Although many of these issues are partially addressed by protocols external to IP such as IPsec, DHCP, NAT, and RSVP/IP CoS, they are by no means the perfect long-term solutions. In addition, NAT, which originally had promise for solving the address depletion issue, has major deficiencies as identified in Section 2.5.7.

Thus there is a need for a new IP, and although the protocol has not yet gained momentum, there is a global overlay now in place (the 6Bone), and vendors are expected to begin to provide wider support in the next year or two. It does solve some major problems in extending IP addressing into the billions of nodes, and its auto-configuration capabilities are quite useful for address management and mobility. It should therefore begin to see wider use toward the beginning of the next decade, with user migration it will be an ongoing process. The lower part of Figure 2.40 depicts the IPv6 header. Major fields include:

Field	Length	Function
Version	4 bits	IP version number—set to "6" for Ipv6
Priority	4 bits	0000–0111—priority for traffic where source provides congestion control
		1000–1111—priority for real-time traffic
Flow Label	24 bits	Uniquely identifies given flow; set by source
Payload Length	16 bits	Length of datagram in octets; max=65,635
		Host must accept min=576
Type	8 bits	Layer 4 protocol (i.e., TCP)
Hop Limit	8 bits	Max=255; decremented at each router

Field	Length	Function
		If "0", then datagram is discarded
Source Address	128 bits	IP source
Destination Address	128 bits	IP destination
Optional Extension	Variable	Fragmentation, authentication, headers, etc.
		Multiple options permitted
		Passed transparently to destination or acted upon at each hop

2.5.2 IP Addressing

The ability of one Internet user to reach any other is based on a global addressing plan at the IP Layer. Every device connected to the Internet is traditionally issued a unique IP address which is carried within the IP header. Routers, forming the core of the Internet, switch the packets based on this header information in much the same way that phone switches establish connections based on global E.164 addressing (telephone numbers).

An IP address is 32 bits (4 octets) in length, offering an address space of over 4 billion unique addresses. There are two address formats:

- Unicast addresses, in the range from 1.0.0.0 to 223.255.255.255, are used for communication between two systems. The earlier practice of splitting this space into Class A, B, and C addresses has gone out of favor with the deployment of *Classless Interdomain Routing (CIDR)* and *supernetting*. The first technique removes the dependence on class boundary routing, while the second aggregates smaller address spaces into larger routable entities. For example, what had been traditionally 256 subnets in the Class B space would now be routed as a single block with the characteristics of a Class A network.
- Multicast addresses, spanning from 224.0.0.0 to 2xx.255.255.255, are associated with *groups*. A user sends traffic to a given group, and any other systems that have subscribed to the group, given administrative constraints, receive the traffic. IP multicasting (or IPmc) is detailed in Section 2.5.4.

To prevent chaos from breaking out, IP addressing is globally administered at the top level by the IANA. Sections of the IP address space have been delegated to regional registries for further allocation. For example, IP networks 62, 193, 195, and 212 have been assigned to RIPE in Europe (`http://www.isi.edu/in-notes/iana/assignments/ipv4-address-space`).

> That all IP addresses are globally unique should probably be taken lightly, since a technique known as *Network Address Translation (NAT)* allows users to translate between globally routable addresses and private IP addresses, some of which may overlap with other users. NAT is detailed in Section 2.5.7.

2.5.3 Routing IP Packets

The existence of global addressing solves only half of the reachability problem. Just as critical is actually being able to *route* the IP packets from one location to another. This is the job of the routers within the Internet, but they somehow have to learn which IP networks, and thus which users, are reachable over which links. Dynamic routing protocols fulfill this role, carrying hierarchical reachability information from router to router across the Internet. Within the Internet, different types of routing protocols are in use, based on whether they are designed for use within a single organization or are more suitable for interconnecting organizations.

The Routing Information Protocol Version 2 (RIPv2) (Malkin, 1994), the Enhanced Interior Gateway Routing Protocol (E-IGRP) (Cisco, 1997), Intermediate System-Intermediate System (IS-IS) (Oran, 1990), and Open Shortest Path First (OSPF) (Moy, 1998) are the best known examples of what are referred to as *interior* routing protocols.

The Border Gateway Protocol Version 4 (BGP-4) (Rekhter, 1995), an *exterior* routing protocol, is used across the Internet backbone to route between organizations and ISPs. RIP is an example of a distance vector protocol, where the path to a distant network is described in terms of *hops*, the number of routers crossed.

In contrast, link-state protocols like OSPF build topology maps within an *area*. The links on this map are described in terms of costs, with packets forwarded on the lowest cost path from source to destination. A routing protocol, depending on its sophistication, may contain features such as load balancing, security, and multiprotocol support.

Figure 2.41 depicts a simple distance vector routing implementation where an IP packet has a choice of two routes to a destination, each with different costs. In this case, the cost parameter is the hop count, and the packet will traverse the lower path, flowing through router 10.10.2.1 and then across the link to 10.20.1.2, a single hop.

FIGURE **2.41.**

Routing

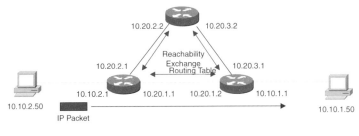

2.5.4 IP Multicasting

An exciting capability just coming into its own across the Internet is the ability to *multicast* data from a single source to multiple destinations. Those who have worked with LAN media will realize that an Ethernet is inherently multicast capable, since traffic (unless switched point-to-point) is sent on a bus to all destinations at once. However, building a scalable multicasting architecture for use across a corporate or ISP backbone is a different matter entirely. It is further complicated by some link layers in use which are either not capable of multicast (Frame Relay) or require additional protocols to utilize their multicasting capability (as is the case with ATM).

In the same way that unicast routing protocols propagate reachability information for unicast destinations, a multicast routing protocol sets up the forwarding path for multicast traffic (Deering, 1989). The most common IPmc routing protocol in use is Protocol Independent Multicast (PIM), whose name is a bit misleading since it is still bound to IP, though not to any specific unicast routing protocol as was the case with earlier IPmc routing protocols (MOSPF, DVMRP).

PIM in fact has two modes of operation as shown in Figure 2.42.

The first, Dense Mode (PIM-DM) (Estrin, 1998a), is more suitable for campus or high-bandwidth environments since it first builds the forwarding tree for all groups in use to all possible destinations. Only then does it prune back the tree from destinations not wishing to receive the traffic, as the figure depicts.

PIM Sparse Mode (PIM-SM) (Estrin, 1998) operates in the opposite way, building the tree on-demand based on who actually wishes to receive the traffic. PIM-SM relies on a Rendezvous Point (RP), a known location where potential receivers of the IPmc data may find the source's transmission. Since this may not be the optimal path to the source, the protocol includes provisions for switching this tree to one rooted on the source, a "source-based

FIGURE **2.42.**

PIM Operation

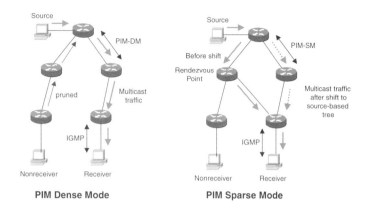

PIM Dense Mode   **PIM Sparse Mode**

tree." It is therefore most useful across a provider's backbone. The two modes can of course interwork. In addition, both rely on Internet Group Management Protocol (IGMP) between the host and first hop router.

PCs wishing to participate in IPmc will have an IGMP capable internetworking stack, allowing it to inform its first-hop router that it wishes to receive traffic belonging to a given group. The router is the entity actually running the multicast routing protocol, forwarding the traffic only onto the local LAN segment if one or more users wish to receive a given group. Users learn of possible IPmc groups by running an application that lists the available sessions.

Of note is an enhancement to multicasting that will prove immensely useful where guaranteed data delivery is a requirement. To date, IPmc has been based on UDP, a nonassured delivery mechanism. This is adequate for video distribution, but it has major shortcomings when trying to guarantee delivery of financial data, for example. A new protocol known as *Pragmatic General Multicast (PGM)* (Speakman, 1998) introduces reliability within the IPmc environment for the first time by building this capability into the routers. PGM should see use beginning in the 1999 timeframe.

In the ADSL environment, IPmc is expected to see widespread use for delivery of streaming video, both within a public service as well as within the corporate, government, or educational environment. A reason for this is the efficiency of IPmc coupled with the data rates available across ADSL. Consider existing web-based video streaming applications at 28.8 or 56 Kbps peak bandwidths. Even 1000 streams fit comfortably into a single FE or OC3/STM1 ATM connection. At the MPEG-1 (1.5 Mbps) rates available across ADSL, a server would be hard pressed to unicast 1000 streams, and the amount of bandwidth required would be unacceptable. There are still some holes to be filled in the ADSL multicasting architecture, however, involving PPP and tunneling. Chapter 4 describes solutions.

2.5.5 Quality of Service

Real-time applications such as VoIP and IPmc-based video streaming expect certain guarantees from the network in terms of bandwidth, data loss, latency, and jitter. Although the demands placed on the network from these IP applications are not as great as the network characteristics expected from the native ATM applications described in Section 2.2, proper network engineering is still a requirement. This is especially true in an ADSL environment where oversubscription is planned as opposed to an exception.

The network architect has a number of tools at his or her disposal at Layer 3. IP Precedence, or *type of service (ToS),* is probably the simplest to implement and involves setting the ToS field in the IP header into one of six service classes. The use of this ToS field is described within the IETF DiffServ architecture (Nichols, 1998 and Blake, 1998). These documents define a "Per Hop Behavior" (PHB) that may be applied against the packets in question (Figure 2.43). A set of PHBs combine into a Per Hop Behavior Group (PHBG). The first PHBG implemented is Assured Forwarding (AF)(Baker, 1999), which divides packets into a maximum of 4 forwarding classes based on precedence, with AF4 taking highest precedence. A networking device will implement this behavior through use of a combination of queuing and buffering mechanisms. One of the best known is Weighted Fair Queuing (WFQ) (Floyd, 1998). As an example, a router may implement the AF PHBG to prioritize VoIP over bulk data—it will assign VoIP to AF4, and all other traffic to AF3 or lower depending on the overall class of the user. To be more specific, users signed up to "gold" service could have their data AF set to 3, while "bronze" subscribers would have AF set to 1.

FIGURE **2.43.**

Quality of Service

When operating across ATM, different IP ToSs could be mapped to different types of ATM PVCs as described in Section 2.2.7. This same field may also be mapped into WRED across routed backbones, while BGP Policy Propagation allows rate-limiting via this field. At the CPE, a feature known as *Committed Access Rate (CAR)* also uses the ToS, for rate limiting.

Less widely implemented is the Resource Reservation Protocol (RSVP) (Braden, 1997), since it requires support at the desktop to be effective. RSVP permits an application to request certain characteristics from the network in terms of bandwidth and delay. If the network can meet this request in the presence of other demands, it will set up the reservation; if it cannot, it won't. A properly designed RSVP implementation is the closest thing available to true application QoS across an IP backbone, but it requires up-front planning. For this reason, initial deployments are within Enterprise networks. In a DSL environment, a situation could be envisioned where applications at the customer site interact with the CPE router, knowledgeable of the bandwidth available across the ADSL uplink and downlink. To be effective, however, requires management of oversubscription in the DSLAM. In this case, the user would probably be mapped to a higher QoS PVC than those subscribers without this capability.

2.5.6 Security (IPSec)

Continuing the discussion of internetworking services, users and network administrators early on identified the need for robust security across a public networking infrastructure. Although many point solutions such as host or link encryption have been deployed, only recently has the Internet community standardized on a single, integrated, multivendor security architecture. This is known as *IP Security (IPSec)* and contains provisions for encryption and user authentication (Thayer, 1997).

IPSec relies on ensuring securing communications within the network infrastructure, which may include the computers connected to the network as well as the routers and switches. All hosts will be equipped with an IPSec *shim* layer, above IP, and although IPSec is independent of any application layer security, it does not preclude this. The architecutre supports multiple encryption algorithms, scalable key management, and the use of digital signatures. This latter feature eliminates the need for two parties to share a "secret" ahead of time.

As an example of the use of IPSec (Figure 2.44), consider a corporation wishing to replace traditional leased lines with connections over the public Internet. Two branch offices (1 and 2) may directly communicate by implementing IPSec, while IPSec capability at the corporate gateway (3) will allow these external hosts to communicate with users internal to the corporate campus. Note that intermediate routers (4) need not have knowledge of the protocol.

FIGURE 2.44.

IPsec

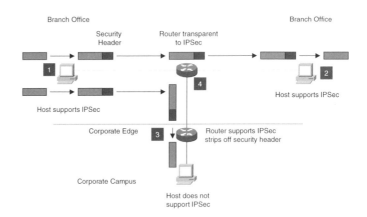

Branch Office

Security Header

Router transparent to IPSec

Branch Office

Host supports IPSec

Host supports IPSec

Corporate Edge

Router supports IPSec strips off security header

Corporate Campus

Host does not support IPSec

> **NOTE**
>
> Although ATM Layer security was not described in Section 2.2, the ATM Forum is developing a comprehensive framework (ATM Forum, 1997c). Deployed at the ATM Layer, it will have many of the features of IPsec while extending it to include additional capabilities such as address-based CUGs. In the future, one would expect deployment of both ATM and IP Layer security.

2.5.7 Network Address Translation

One technique sometimes useful in conserving IP address space is known as *Network Address Translation (NAT)* (Egevang, 1994). Here, a gateway converts between a corporation's private address space to globally routable addresses. For example, a large corporation may have been assigned the equivalent of only 64,000 IP addresses (a single Class B network, or 256 subnets given 8 bit subnet addressing). This is obviously not sufficient for everyone within the corporation to reach the Internet. Or, a corporation may wish to provide a local subnet to every telecommuter. NAT allows the corporation to assign private address spaces, remapped into public addresses when the traffic crosses the corporate boundary.

Within the Internet community, three networks, 10/8, 172.16/12, and 192.168/16 (Rekhter, 1996), have been designated as unroutable and should be used if possible for private addressing. Figure 2.45 depicts this in greater detail. Here, a host internal to the corporate Intranet, 10.10.15.7, wishes to communicate with a host on the Internet. The NAT maintains an address pool (118.32.10.1–118.32.10.255) which is used to remap the host's internal address to one routable over the Internet. Then 10.10.15.7 is remapped to 118.32.10.12. In the reverse direction, packets addressed to this host are mapped back into 10.10.15.7.

Figure 2.45.

Network Address Translation

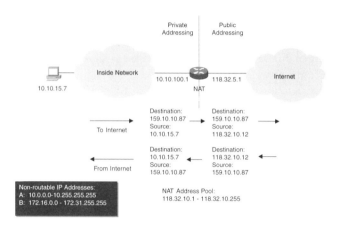

NAT does present problems, however, in terms of IPsec and reachability. One concern is that IPsec assumes an association between an end system and an IP address. If this address is altered within the network, the mechanisms fail. Even more problematic is reachability for multimedia applications such as Microsoft's NetMeeting. Due to the remapping, and in the absence of sophisticated directory systems, a user behind a NAT is effectively unreachable. To overcome such a limitation whereby incoming traffic won't be "NATted" to the proper destination, there are solutions available implementing Proxy functions for Multimedia applications (e.g., Proxy Gatekeeper for H.323 applications); usually this proxy element is coresident with a border router or firewall, and therefore it is visible from both the public Internet and the private Enterprise. The role of the proxy is to present to the public Internet a high-layer identifier of an internal destination (e.g., the H.323-ID) and then map the incoming traffic to the proper internal private address: Note that these are IP addresses as opposed to the H.323-ID which has an application layer meaning

2.5.8 The Dynamic Host Configuration Protocol

The Dynamic Host Configuration Protocol (DHCP) (Droms, 1997) allows a PC (or router for that matter) to obtain an address dynamically from a DHCP server. This is almost a requirement in the telecommuting environment, where users first obtain an IP address in the office for their laptops, and then login from home. In an ADSL environment, when a subscriber connects to a given upstream service, be it ISP or corporate gateway, a DHCP server will supply the address.

A related protocol is the IP Configuration Protocol (IPCP), used many times to provide addresses to CPE routers. Here, the subscriber's router acts as the DHCP server, while the router obtains a device for its upstream interface via IPCP. Usually, NAT is used to map the DHCP-assigned addresses into the corporate, globally routable address space. An older protocol, though one still in use, is BOOTP, on which many of the DHCP concepts are based. Once major difference is the concept of leases within DHCP, where a user is assigned an address for a set period of time. This is useful for address optimization and if many users are mobile.

Figure 2.46 depicts the sequence of events under DHCP operation. Note that if there is a router in the path, it must support an IP helper-address.

FIGURE 2.46.

*Dynamic Host
Configuration Protocol*

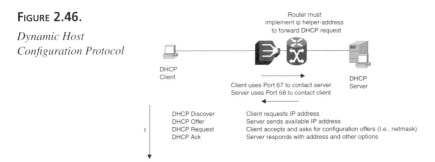

2.5.9 The Domain Name System

Related to IP addressing is the Domain Name System (DNS) (Mockepetris, 1987), a hierarchical set of servers deployed across the Internet providing an address to name resolution. Consider the hostname www.foo.com, which actually maps to the IP address 192.31.7.130. If users always had to enter IP addresses, the Internet would be far less user friendly. This is especially true for Web and e-mail traffic. The use of DHCP would all but be impossible, since the user's address changes over time—thus the need for the DNS and a tightly regulated system for domain assignment.

DNS is based on ITU-designated country codes (.us, .uk, .au, etc.) corresponding to top level domains administered within each country and a set of additional top level domains (.com, .gov, .edu, etc.) handled by the InterNIC. All but .com apply to the United States only. When a user requests a connection to a host such as www.foo.com, the PC issues a request to the default DNS, unless the mapping is already cached. The DNS will either respond or relay the request up the hierarchy until it reaches a server able to respond.

2.5.10 Tunneling

Finally, before leaving Layer 3, a brief introduction of Internet tunneling is in order. Traditionally, users create VPNs at Layer 2 using leased lines, Frame Relay DLCIs, and more recently, ATM VCs. These techniques all suffer in terms of scalability and manageability, especially when creating large full or partial meshes. An added disadvantage is the decoupling of the Link Layer (Frame Relay or ATM) and Layer 3, requiring traffic in many instances to jump in and out of the transport network on its way to a destination. Native Layer 3 techniques eliminate this discontinuity, and in the ADSL space, the most widely deployed protocol is expected to be the Layer 2 Tunneling Protocol (L2TP) (Valencia, 1998). This allows a VPN tunnel based on PPP to be created between two Layer 3 entities, through which traffic (even non-IP) may flow, with a user no longer bound to a single destination or IP address.

In an ADSL deployment, L2TP tunnels are usually created between an Aggregator functioning as a L2TP Access Concentrator (LAC) and an ISP or corporate router performing a L2TP Network Server (LNS) function. Figure 2.47 depicts the sequence of events in establishing an L2F or L2TP tunnel.

FIGURE 2.47.

Layer Two Tunneling Protocol (L2TP) and Layer Two Forwarding (L2F)

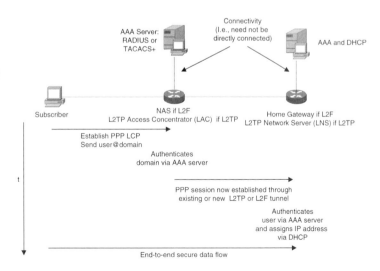

A user establishing an L2TP tunnel will first establish a PPP link to the LAC. This device (the aggregator in the case of ADSL) uses the user's fully qualified domain name (user@domain) to query an AAA server or a local configuration file as to the proper L2TP tunnel to use. The PPP session is now encapsulated in L2TP for transport to the destination LNS. Here, final authentication takes place at the destination AAA server, and the user is assigned an IP address via DHCP. In some instances, an end-to-end tunnel may be created between the CPE and the router, though this requires additional overhead at the CPE and does not allow one to combine traffic from multiple users into a single tunnel. L2TP plays a major role in one of the end-to-end data models described in Chapter 4 (and in the examples).

Other tunneling protocols expected to see use include the Layer 2 Forwarding Protocol (L2F) and the Point-to-Point Tunneling Protocol (PPTP). A separate set of tunneling techniques based on MPLS, Tag-VPNs, or even IPsec should also see wide use. As these tunneling techniques are expected to play a major role in enabling advanced services across ADSL, they are covered in detail within the end-to-end service models in Chapter 4.

2.5.11 The Transmission Control Protocol and the User Datagram Protocol

The Transport Layer, or Layer 4, provides delivery of packets between two end-systems. Over IP, the two transports in place are the Transmission Control Protocol (TCP) (Postel, 1981a), providing assured delivery by establishing a session between the two systems, and the User Datagram Protocol (UDP), described in the next section.

TCP provides the end-to-end reliability that does not exist at the connectionless IP Layer and includes provisions for flow control, multiplexing of user sessions, connection establishment, and even precedence and security. Most important, the protocol includes mechanisms for backing off data transmission in times of congestion, and retransmitting the data if packets are lost. It accomplishes this by buffering a certain amount of data transmitted (the window) and waiting for an acknowledgment from the destination stating that the next set of data may be sent. If no acknowledgment is received, the sender will retransmit the data. In addition, lost packets signal that the network is experiencing congestion, at which time TCP reduces its transmission speed. Figure 2.48 depicts the TCP header, along with an example of the windowing mechanism.

2

FIGURE **2.48.**

TCP

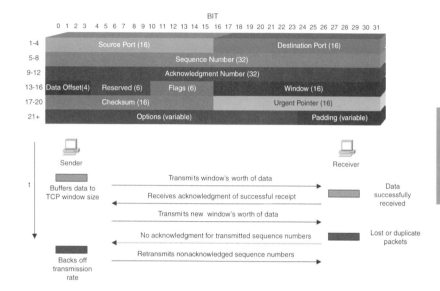

Major fields in the TCP header include:

Field	Length	Function
Source Port	16 bits	Identifies unique user session May be well known
Destination Port	16 bits	Identifies unique user session
Sequence Number	32 bits	# of first data octet in segment
Acknowledgment	32 bits	Next sequence number <sender>Number is expecting to receive
Window	16 bits	Number of data octets sender is willing to buffer

As introduced in the section on ATM, one concern is the interaction between the Transport and ATM Layers in preserving data *goodput* across the network. Goodput is the data that is actually usable by the destination (arriving in one piece), as opposed to throughput which is the data thrown into the network by the source. If an ATM network (or any network for that matter) is experiencing congestion, resulting in lost data either intentionally (via RED) or unintentionally, TCP will realize that it is transmitting too fast and will backoff the transmission rate (known as *slow-start*). It must then be capable of retransmitting the lost data, hopefully stored in its buffers. The requirement, then, is to provide buffering equal to the feedback loop time and bandwidth. This is known as the *Bandwidth * Round-trip-time* *(bw*rtt) product.* The same amount of buffering is also required on ATM switch interfaces if implementing congestion avoidance such as ABR.

One problem identified is TCP's propensity to probe for the highest possible bandwidth. If multiple sessions do this in parallel, they will all experience congestion, with TCP backing off to a lower transmission speed. The process is then repeated, resulting in a sawtooth goodput pattern. The solution is to insure that sessions do not synchronize, either through the use of RED or WRED across packet backbones, or via EPD/TPD within an ATM switch.

In contrast to TCP, the User Datagram Protocol (UDP) doesn't rely on session establishment and thus is incapable of providing any guarantees on delivery. It is most useful for one-shot transactions, although it is also used for IPmc since one-to-many session establishment is not a trivial problem to solve—thus the PGM described earlier.

Summary

Beginning with a brief history of ADSL standardization, this chapter then detailed the two encodings in use—CAP and DMT. It then built upon this ADSL layer by focusing on ATM as the data, voice, and video encapsulation of choice across the local loop. This led to an explanation of the various ATM data models, and finally the higher layer networking protocols. Obviously, an ADSL deployment is rather complex in the protocols and standards deployed at the various layers and among the different components.

Whereas this chapter focused on the protocol architecture forming the basis for end-to-end service deployments, the next chapter focuses on the physical architecture, service components, and management. These two chapters then form the basis for the end-to-end service models described in Chapter 4 and the implementation examples in Chapter 5.

Endnotes

ADSL Forum, Framing and Encapsulation Standards for ADSL: Packet Mode, ADSL Forum TR-003, June 1997.

ADSL Forum, Default VPI/VCI Addresses for FUNI Mode Transport: Packet Mode, ADSL Forum TR-008, March 1998.

ADSL Forum, An End-to-End Packet Mode Architecture With Tunneling and Service Selection, ADSL Forum TR-011, June 1998.

ADSL Forum, Network Management—Working Text, ADSL Forum WT-024, June 1998.

ANSI, Rate Adaptive Asymmetric Digital Subscriber Line (RADSL) Metallic Interface, Draft ANSI Specification Revision 2a, October 1997.

ANSI, Standards Project for Interfaces Relating to Carrier to Customer Connection of Asymmetrical Digital Subscriber Line (ADSL) Equipment, ANSI T1.413 Issue 2, June 1998.

ATM Forum, ATM Forum Traffic Management Specification Version 4.0, AF-TM-0056, ATM Forum, April 1996.

ATM Forum, ATM User Network Interface (UNI) Specification Version 4.0, AF-SIG-0061.000, ATM Forum, July 1996.

ATM Forum, Private Network–Network Interface Specification Version 1.0, AF-PNNI-0055, ATM Forum, March 1996.

ATM Forum, Circuit Emulation Service Interoperability Specification Version 2.0, af-vtoa-0078.000, ATM Forum, January 1997.

ATM Forum, Audio/Visual Multimedia Services: Video on Demand v1.1, af-saa-0049.001, ATM Forum, March 1997.

ATM Forum, Frame Based User-To-Network Interface (FUNI) Specification v2.0, AF-SAA-0088.000, ATM Forum, July 1997.

ATM Forum, LAN Emulation Over ATM Version 2—LUNI Specification, AF-LANE-0084.000, ATM Forum, July 1997.

ATM Forum, Multi-Protocol Over ATM Specification v1.0, AF-MPOA-0087.000, ATM Forum, July 1997.

ATM Forum, Voice and Telephony over ATM—ATM Trunking Using AAL1 for Narrowband Services, af-vtoa-0089.000, ATM Forum, July 1997.

ATM Forum, ATM Security Specification Version 1.0, btd-security-01.04, ATM Forum, September 1997.

Baker, F., Heinanen, J., and Wroclawski, J., Assured Forwarding PHB Group <draft-ietf-diffserv-af-04.txt>, January 1999.

Blake, S., et al., RFC-2475: An Architecture for Differentiated Services, December 1998.

Braden, R. et al., RFC-2205: Resource ReSerVation Protocol (RSVP)—Version 1 Functional Specification, September 1997.

Callon, R. et al., A Framework for Multiprotocol Label Switching, Internet Draft, draft-ietf-mpls-framework-02.txt.

2

Cisco, Introduction to Enhanced IGRP (EIGRP), Cisco Systems,
`http://www.cisco.com/warp/public/459/7.html`.

Deering, S., RFC 1112: Host extensions for IP multicasting, August 1989

Deering, S., and Hinden, R., RFC-1883: Internet Protocol, Version 6 (IPv6) Specification,
December 1995.

Droms, R., RFC-2131: Dynamic Host Configuration Protocol, March 1997.

Egevang, K., and Francis, P., RFC-1631: The IP Network Address Translator (NAT), May
1994.

Estrin, D., et al., RFC 2362: Protocol Independent Multicast—Sparse Mode (PIM-SM):
Protocol Specification, June 1998.

Estrin, D., et al., Protocol Independent Multicast Version 2 Dense Mode Specification,
Internet Draft, draft-ietf-pim-v2-dm-00.txt, August 1998.

Floyd, URL to RED references: `http://www-nrg.ee.lbl.gov/floyd/red.html`

Gross, G. et al., RFC-2364: PPP over AAL5, July 1998.

Heinanen, J., RFC 1483: Multiprotocol Encapsulation over ATM Adaptation

Layer 5, July 1993.

ITU-T, Transport of MPEG-2 Constant Bit Rate Television Signals in B-ISDN, ITU-T J.82,
July 1996

ITU-T, Asymmetrical Digital Subscriber Line (ADSL) Transceivers, Draft, ITU-T Recommendation G.992.1 (ex: G.dmt), October 1998.

ITU-T, Handshake procedures for Digital Subscriber Line (DSL) transceivers, Draft ITU-T
Recommendation G.994.1, October 1998.

ITU-T, Test Procedures for Digital Subscriber Line (DSL) Transceivers, Draft ITU-T Recommendation G.996.1, October 1998.

Laubach, M, and Halpern, J., RFC-2225: Classical IP and ARP over ATM, April 1998.

Maher, M., RFC 2331: ATM Signalling Support for IP over ATM—UNI Signalling 4.0
Update, April 1998.

Malkin, G., RFC-1723: RIP Version 2—Carrying Additional Information, November 1994.

McGregor, G., RFC-1332: The PPP Internet Protocol Control Protocol (IPCP), May 1992.

Mockapetris, P., RFC-1034: Domain Names—Concepts and Facilities, November 1987.

Moy, J., OSPF Version 2, April 1998.

Nichols, K. Blake, S., Baker, F., and Black, D., RFC-2474: Definition of the Differentiated Services Field (DS Field) in the IPv4 and IPv6 Headers, December 1998.

Oran, D., RFC-1142: OSI IS-IS Intra-domain Routing Protocol, February 1990.

Postel, J., RFC-768: User Datagram Protocol, August 1980.

Postel, J., RFC-791: Internet Protocol, IETF, September 1981.

Postel, J., RFC-793: Transmission Control Protocol, September 1981.

Rekhter, Y., and Li, T., RFC-1771: A Border Gateway Protocol 4 (BGP-4), March 1995.

Rekhter, Y. et al., RFC-1918: Address Allocation for Private Internets, February 1996.

Rekhter, Y. et al., RFC-2105: Cisco Systems' Tag Switching Architecture Overview, February 1997.

Simpson, W., ed., RFC-1662: PPP in HDLC-like Framing, July 1994.

Speakman, T., Farinacci, D., Lin, S., and Tweedly, A., PGM Reliable Transport Protocol Specification, Internet Draft, draft-speakman-pgm-spec-02.txt, August 1998.

Thayer, R., Doraswamy, N., and Glenn, R., IP Security Document Roadmap,

Internet Draft, draft-ietf-ipsec-doc-roadmap-02.txt.

Valencia, A., Layer Two Tunneling Protocol—L2TP, Internet Draft, draft-ietf-pppext-l2tp-11.txt, May 1998.

2

Chapter 3

ADSL Infrastructure

Building upon the protocols and standards described in Chapter 2, this chapter details the end-to-end ADSL infrastructure. First, the next generation Internet architecture is introduced. Next, the various components of this architecture are described in further detail, including the Customer Premise Equipment, the Central Office, service aggregation, access networks, Digital Loop Carrier environments, and corporate and ISP gateways. The next areas addressed are the various server components within the architecture, including content, caching, and provisioning. Finally, the effect of the regulatory environment on hardware deployment is discussed.

3.1 The New Internet Infrastrucure

Until recently, the Internet infrastructure was composed of a backbone infrastructure mainly based on T3/E3 and OC3/STM1 links. Access methods to the Internet include leased lines and Frame Relay lines up to T1/E1 or T3/E3, modem, or ISDN connectivity.

The new infrastructure is based on multiple fiber links in the backbone, supporting ATM, SONET/SDH, or DWDM transmissions and speeds from OC3/STM1 to OC48/STM16 and even OC192. The IP protocol may be carried over an ATM infrastructure (both overlay and via MPLS mechanisms) or layered directly over SONET/SDH or DWDM. At the same time, high-speed access technologies such as HFC supporting cable modems, high-speed leased lines, DSL, and wireless are emerging or are being deployed.

Translating this next generation architecture into deployable reality (Figure 3.1), the customer premises include combinations of ADSL modems, PCs, phonesets, and settop boxes.

A typical configuration (1) will consist of a PC connected to an external ADSL router, feeding through a POTS/ISDN splitter to the local loop. The router may support voice over IP (VoIP) or voice over ATM (VoATM) depending on the application. A simpler architecture (2) consists of a PC connected to an external ADSL bridge. An alternative is to deploy a settop box (3) with an integrated ADSL modem. The first element encountered within the central office (CO) is the Main Distribution Frame (MDF) (4), which may include POTS/ISDN splitters. Here, the POTS/ISDN voice traffic is forwarded to the existing voice switch (8). The copper pairs then extend to the Digital Subscriber Line Access Multiplexer (DSLAM) (5) which contains the actual ADSL modems. Depending on the service profile of the ADSL provider, the DSLAM may connect to a colocated service aggregator (7) or may forward all traffic across an ATM access network (6) to a remote aggregator. The DSLAM may also include the capability of actually terminating the POTS/ISDN voice traffic, forwarding the traffic across a T1/E1 trunk to the voice switch.

The service aggregator, either colocated with the DSLAM or in a Layer 3 point of presence (PoP), includes interfaces to various application servers and web caches (10), as well as connecting to a local or remote AAA server. It may also connect to a VoIP gateway (9) for for PSTN interworking. Traffic leaving the aggregator flows across a regional ATM network (11) or packet network (12), finally terminating at corporate or ISP gateway sites (13).

In those locations where Digital Loop Carrier (DLC) systems have been deployed and the provider wishes to offer ADSL service, a DSLAM will be located within the DLC enclosure in the subscriber's neighborhood (14), and traffic will be forwarded directly to the service aggregator or across an ATM access network.

These basic ADSL architecture components play a vital role in the access layer of the next generation Internet infrastructure. Consider a traditional access environment of dial servers, both POTS and ISDN, connected to routers, which would then natively interconnect across a MAN or WAN or, alternatively, use ATM or Frame Relay for interconnectivity. Some high-end customers would connect via leased lines, Frame Relay, or even ATM. The service provider would also deploy application servers offering DNS, e-mail, and Web, depending on the sophistication of the service offering. Fast forward to the end of the 1990s, with cable modems, the various flavors of DSL (IDSL, SDSL, ADSL, VDSL, HDSL), and even wireless. The service infrastructure has evolved into a hierarchy of web caches and re-directors,

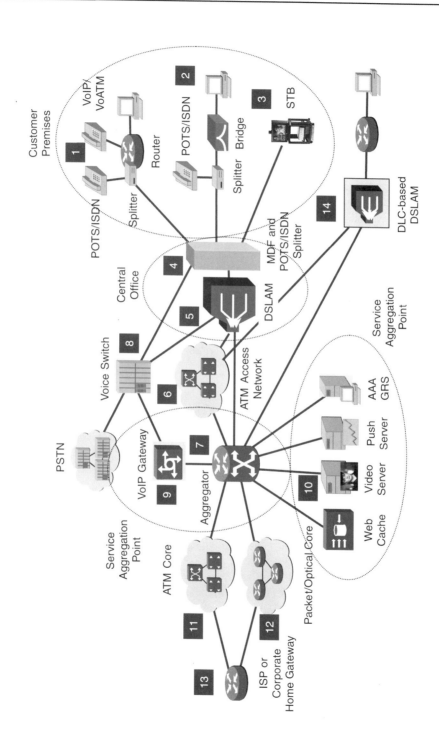

Figure 3.1.

ADSL End-to-end Architecture

3

streaming video servers, and sophisticated user provisioning systems combining authentication, accounting, quality of service (QoS) assignment, directory services, and even network
modeling. Internet access now scales from tens or hundreds of thousands of users to tens of
millions. Multimedia applications are commonplace, and Layer 3 VPNs scale secure connectivity. The sections that follow detail the basic hardware components introduced in Figure 3.1 and accompanying descriptions. Later, this chapter describes service architectures
based on this infrastructure.

3.2 The Customer Premises

Within an ADSL deployment, the customer premises, either business or home, may at times
be almost as complex as the CO. This environment contains the user's PC or Settop Box,
also known as *Terminal Equipment* in the parlance of the ADSL Forum. These devices connect to a Premises Distribution Network (PDN), over which the data accesses the ADSL
modem, or ATU-R. This then connects to a POTS/ISDN splitter and finally to the local loop.
The ADSL Forum formalizes this into a Customer Premises Equipment (CPE) reference
model consisting of a set of logical entities and interfaces. Figure 3.2 depicts the architecture of the customer premises portion of the ADSL environment.

FIGURE **3.2.**

*ADSL Forum Reference
Model*

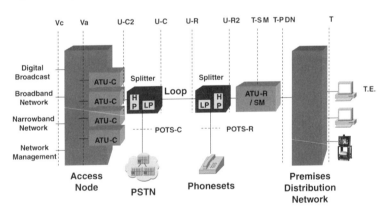

Moving in the direction of the loop to the user, the first interface encountered is the U-R
between the copper loop and the splitter. This splitter presents a POTS-R interface to the
phonesets and a U-R2 interface to the ATU-R. Normally, the connection between the splitter
and both the loop and the phoneset will be RJ11, while that to the ATU-R will be RJ14 (a 4
wire RJ11). Although the reference model identifies a T-SM interface between the ATU-R
and an entity called the *Service Module (SM)*, in almost all implementations, the ATU-R and
SM are one in the same. However, if these two devices are separated, the interconnect may

be one of a number of interface types, with RJ45 (implementing V.11) being the most common. Other interface types include TIA-530, X.21, V.24, and X.24.

The SM component now presents a T-PDN (Premises Distribution Network) interface to the in-building Ethernet, ATM25, Universal Serial Bus (USB), or even an IEEE-1394 network. Although most first generation ATM/ADSL ATU-Rs were equipped with ATM25, Ethernet and USB are expected to become more common over time. Serial interfaces such as TIA-530, T1, or E1 are also possible, although these have not been implemented since they see little use in in-building installations.

Terminal Equipment (TE), which may include PCs and Settop Boxes, connect to this PDN. Although this terminology may seem complex, it is useful in allowing implementers to standardize on interfaces and functionality, and is all but invisible to the end user who goes out and purchases a broadband PC with integrated ADSL NIC and plugs it into the phone jack on the wall. Some or all of these components may in fact be combined. For example, a PC with an integrated ADSL modem has no need for a PDN.

The following sections look at each of these components in detail, following the signal flow from the user to the ADSL loop, and then into the CO.

3.2.1 Terminal Equipment

Terminal Equipment is a term that seems to appear in every reference model and standard, a technical way of referring to a computer (PC, Mac, or UNIX workstation) or settop box. In most business and home telecommuter ADSL deployments, PCs will predominate, either connected to the ATU-R via an in-building network (described in section 3.2.2) or where permitted and desirable, integrating the ATU-R functionality on a NIC. An alternative architecture, more suited for some casual Internet users, is to deploy a device known as a settop box. However, this unit is not the type of system associated with cable or DBS networks. It is more akin to a device integrating PC and browser capability into a low-cost package, relying on the existing TV for display. In some markets where MPEG-2 video is encapsulated directly over ATM across the ADSL loop, you may also expect deployment of a more classic settop box providing video decoding. Note that nothing precludes integrating this function within the device, and almost every settop box expected to be deployed will have at least some Internet/browser capability.

Looking back at the PC in greater detail, it will contain an internetworking stack capable of supporting the data encapsulation of choice and dependent upon the choice of technology for the in-building network. The simplest encapsulation, and that most analogous to the existing Internet dial-up model, is the Point-to-Point Protocol (PPP) as introduced in Chapter 2. The dial-up model is the actual architecture used in today's services when a user connects to his Internet service provider (ISP) via modem or via ISDN: The extension of this model to ADSL line, which is an always-on type of connectivity, will allow a smooth migration of users and services from a narrowband to a broadband solution. In this model, a user initiates a dial-up session (via Microsoft Networking) and enters a user name (fully qualified with domain if required) and a password.

3.2.2 Premises Distribution Network

Within the context of the ADSL reference model, the Premises Distribution Network (PDN) is the technology connecting the user's terminal equipment (PC or STB) with the ADSL modem (ATU-R), which is described in section 3.2.3. This PDN may take many forms, including Ethernet (including Fast Ethernet), ATM, USB, or even IEEE-1394 or the HomePNA in the future. The choice will depend on the user's mix of services, the ADSL provider's choice of ATU-R interfaces, and the regulatory environment. In most Internet-centric ADSL deployments, Ethernet will probably be the service interface of choice, connecting the ATU-R with the user's PC, hub, or LAN switch. This architecture will apply both where the ADSL provider terminates at service at an NT interface, as well as in locations where the customer owns the CPE terminating the service. Ethernet has advantages in that it is the most common technology within the business environment, although mapping PPP is more complex than with ATM.

3.2.2.1 Asynchronous Transfer Mode (ATM)

ATM is expected to predominate (or at least play a major role) in those countries where the ADSL provider owns the network termination (that is, most European providers). In fact, with the introduction of the ATM-centric ADSL delivery model, it was envisioned that ATM25 would be *the* interface of choice. Although this is no longer the choice, most providers still specify ATM as one of the two required interfaces, the other being Ethernet. Also, the preferred interface into STBs is still ATM, although the eventual penetration of these VoD-centric devices is still in doubt. In the United States at least, it now looks as if ATM25 will not be the preferred service interface. In the vast majority of cases, the CPE presents Ethernet to the customer. Note that in deployments where the user's PC includes the ADSL modem (common in the residence), the PDN does not exist (unless one might consider the PC's bus to assume the role of the PDN).

Analyzing ATM and Ethernet from their capability to support application QoS requirements (for example, within video distribution), ATM has always been promoted as the technology more capable of handling the bandwidth and delay requirements of video streaming. Given an ATM-connected video server at the content provider and direct encapsulation of the video stream across ATM (i.e., MPEG2 over AAL5), this may be true. However, the majority of video delivery within the expected ADSL application space is IP-based, many times making use of IP multicasting. Here, the choice of delivery technologies is less relevant, witnessed by the success of video over Ethernet within the enterprise. Near-term efforts to better integrate Layer 3 QoS with ATM (described later) will only help things. Therefore, the need to support video cannot be used as a decision criteria in selecting the PDN technology.

3.2.2.2 Ethernet

The Ethernet interface will comply with either Ethernet Version 2.0 or ANSI/IEEE 802.3, presenting 10BaseT on an RJ-45 connector, while ATM25 will comply with the ATM Forum's recommendation. In both cases, UTP5 would extend from the ATU-R to the PC.

Most ISPs and SPs have recognized the need to extend the PPP residential model to a greater set of transport technologies: Currently PPP runs on top of dialup, ISDN, and ATM links while Ethernet will be used mostly by customers of ADSL service. For this reason there are multiple activities in the industry and within the IETF to extend PPP connectivity over Ethernet, namely Broadband Modem Access Protocol (BMAP) proposed by Intel and PPP over Ethernet (PPPoE). Both of these are detailed in Chapter 4.

3.2.2.3 Universal Serial Bus (USB)

The relative newcomer is the Universal Serial Bus (USB), supported by almost every PC manufacturer, and is in fact native to more PCs and laptops than is Ethernet. This point is important, in that the majority of consumer (as opposed to office) PCs do not natively support Ethernet—this requires an additional interface. Thus USB could be the ideal data interface within the household. The standard defines both data pairs and power, with a maximum throughput of 12 Mbps at 5 meters. Thus it is more suitable for short-range interconnects rather than an in-building distribution network. Devices connect directly to the PC, known as a *host* or *network master*, or via hubs to the PC. In both cases, they appear as logically connected to the PC. The USB supports up to 127 devices and will see greatest use in connecting low- to medium-speed devices such as printers, phones, scanners, and modems to PCs. A USB-equipped PC will connect to the ADSL modem in the same way as with Ethernet, and, as described in Chapter 4, the requirement for PPP session mapping is about the same.

3.2.2.4 IEEE-1394

Looking a bit further afield, IEEE-1394 has also been proposed as a service interface, although very few PCs support this at present. IEEE-1394, also known as Firewire, is another serial interface making its way into computer equipment, although it is already available on a variety of consumer electronics devices such as digital video cameras. Firewire is more versatile than USB at the cost of additional complexity and is designed to replace parallel and serial interfaces and even the SCSI. The interface requires special cabling containing 2 data pairs and power, with a typical reach of 4.5 meters and a bandwidth of 100, 200, or 400 Mbps (with even Gbps rates predicted for the future). As with the USB, this limits IEEE-1394 to local interconnect. The standard supports up to 1023 buses, each with 63 nodes, although it does not require a master. In contrast to the USB, uses of 1394 will include interconnecting digital TVs, camcorders, STBs, and digital cameras, an area in which it is especially suitable due to its support for real-time data. On the negative side, the mapping of user data into the technology is not at the top of the priority list within the ADSL Forum. With saying that, the IETF is in fact standardizing the mapping of IP across IEEE-1394.

- 3.2.2.5 Home Phoneline Networking Alliance (HomePNA)

A recent development is the Home Phoneline Networking Alliance (HomePNA), proposing a 1 Mbps CSMA/CD-based network over the existing home phone wiring. This architecture has been embraced by a number of the consumer PC and software vendors as a low-cost in-house network to connect PCs and peripherals to a single Internet point of presence. Although the first iteration of the standard is at 1 Mbps, based on the Tut HomeRUN® technology, future versions may operate at 10 or even 100 Mbps. Technologies such as USB and IEEE-1394, more suitable for a single room, may connect to a HomePNA backbone. One potential problem with the architecture is compatibility with future VDSL deployments due to spectrum overlap, although this does not detract from the near-term desirability of the proposal.

3.2.2.6 Future Interface Possibilities

Other PDN interfaces are of course possible, predicated upon support of the ADSL CPE vendors. For example, a modular router with an integrated ADSL modem could in fact present a wide variety of interfaces. Some examples include Token Ring and FDDI. In the final analysis, the choice of PDN technology almost falls out of the equation when looking at the higher layer end-to-end architecture, except in some niche applications (such as the ATM-based VoD described previously in this section). Table 3.1 summarizes the most common PDN alternatives in terms of capabilities. Note that ATM25, Ethernet, and the HomePDN are suitable for home networking, while the USB and IEEE-1394 standards are local interconnects (i.e., reach would not exceed that of a home office).

TABLE 3.1. PREMISES DISTRIBUTION NETWORK ARCHITECTURES

	ATM25	Ethernet	HomePDN	USB	IEEE-1394
Bandwidth	25 Mbps*	10/100 Mbps	1 Mbps (10 Mbps+ future)	12 Mbps (half-duplex)	400 Mbps (half-duplex)
Reach	50m (Cat5)	100m (Cat3)	500 feet (in-home phone wiring)	5m (shielded)	4.6m (shielded)
Installation	PC NIC	Most PCs require NIC	PC NIC; motherboard in future	Included with PCs	Included with PCs
Availaiblity	No laptop NICs	Wide	1H99	On new PCs/ laptops	On some PCs/laptops
Multiple Host Support	Requires switch (high cost)	Hub (low cost)	NativeHub (low cost)	Hub	
EMC	Good	Poor	Problem with VDSL due to frequency spectrum	Good due to limited reach	Good due to limited reach
QoS	Good	Minimal	Minimal	Good	Good
Protocol Support	Native ATM, PPP, 1483	PPP via BMAP/ PPPoE, tunneling; bridging	PPP via PPPoE possible; bridging	PPP via BMAP	PPP via BMAP
SVC support	Yes	Via ATU-R	Via ATU-R	Via ATU-R	Via ATU-R

*Throughput to ADSL loop is of course limited to ADSL data rate.

3.2.3 ATU-R (ATU-NT)

Working inward from the user to the network core, after the PDN (Figure 3.2), is the CPE, the device that connects the user's PC, workstation, or LAN switch to the copper pair. This CPE may take many forms depending on the type of user, the service offering, technology preference, and the regulatory and business environment. Figure 3.3 depicts these two environments, the left column detailing passive termination options, where the service provider does not control the CPE, and the right side detailing active terminations, where the provider installs an active termination point at the customer premises.

FIGURE **3.3.**

CPE Options

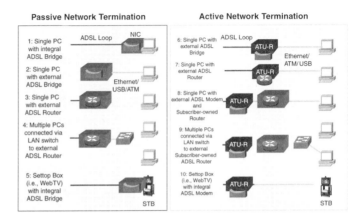

Table 3.2 summarizes the various user and provider-owned data CPE options. Chapter 4 looks more closely at the different bridged and routed data models.

TABLE 3.2.   CPE OPTIONS

	User owns ADSL Modem	*Service provider owns ADSL modem*
Residential	PC with ADSL modem unless provider requires external ATU-R for management	PC with ATM25, Ethernet, or USB connected to external Layer 2 (bridging) ATU-R
Single Corporate Telecommuter	PC with Ethernet connected to external Layer 3 (routing) ATU-R (although a Layer 2 ATU-R could suffice based on requirements)	PC with Ethernet or ATM25 connected to external Layer 2 or Layer 3 ATU-R depending upon requirements.
Multiuser Small Office/Home Office	PC with Ethernet connected to LAN switch or hub, which then connects to user-owned Layer 3 ATU-R (if managed Layer 3 service, operator may own/operate Layer 3 ATU-R)	PC with Ethernet connected to LAN switch or hub, which then connects to operator-owned Layer 3 ATU-R

3.2.3.1 Passive Network Termination

Within the passive network termination environment, common in the United States, there are multiple connectivity options. The simplest is via an integral PC NIC (Scenario 1 in Figure 3.3) using Windows 95/98 plug-and-play drivers. Note that nothing here mandates a discrete PC NIC. The ADSL modem functionality may be integrated within the PC motherboard, allowing the PC to connect directly to the copper pair. This latter option is common under G.lite, although PCs are expected to be equipped with full-rate ADSL as well. In a single PC environment, more sophisticated options include external ADSL-capable bridges (Scenario 2) and routers (Scenario 3).

The choice will depend on the service offering in question. If operating as a bridge, the ATU-R will contain no Layer 3 intelligence, simply remapping and then segmenting the user's preferred PDN data encapsulation onto ATM. For example, in the case of bridged Ethernet, the ATU-R will remap the user's Ethernet data encapsulation into RFC-1483 bridging across ATM. The ATU-R is more complex, though more versatile, if a router. Here, it not only performs re-encapsulation of the user's data but also segments the user's LAN domain from that of the ADSL provider. This is useful in providing a clean demarcation at Layer 3, helps avoid propagating bridged background traffic across the ADSL link, and is a reason why routers are the devices of choice for branch offices and telecommuters. Routers also support value-added features such as encryption, compression, protocol tunneling, DHCP relay, and even VoIP. In the vast majority of multiuser sites, this is the anticipated option due to issues in managing the local address space. Here, the user's PC would be equipped with an ATM25, Ethernet, or USB interface to the external router or bridge. In the SOHO or branch office environment, multiple PCs could be connected via a hub or LAN switch (Scenario 4). Finally, the ADSL modem capability may be included within a Settop box (Scenario 5).

3.2.3.2 Active Network Termination

Now looking at active network termination, the simplest architecture is where the subscriber's PC connects to an external ADSL modem operating in bridge mode (Scenario 6). This connection may be via Ethernet, ATM25, or USB. A more sophisticated CPE model introduces routing into the ADSL modem (Scenario 7). In both of these instances, the provider owns and operates the ATU-R. Alternatively, the subscriber may own a router connected to a provider-owned ATU-R (Scenario 8). Adding a hub or LAN switch (Scenario 9) allows multiple PCs to connect to the service. Paralleling Scenario 5, active termination also supports Settop boxes (Scenario 10) via an external ATU-R.

The real question, then, is when should a user select Layer 2 or Layer 3 termination? This depends on the type of ADSL service offering and if telecommuting is a factor of corporate policy. The casual Internet user, in most cases with a single PC, will usually require no local Layer 3 capability. In contrast, a telecommuter may operate multiple devices behind the termination point, each with its own IP address. Here, a router equipped with Network Address Translation (NAT) will prove useful. From the perspective of the corporation, a router allows more sophisticated user security and address management. This model holds even if the ADSL provider is offering end-to-end ATM virtual circuit connections (VCCs) between the user and the corporate gateway.

Now looking at the broader picture, the small business or corporate branch office will frequently require additional value-added services such as VPNs, firewalling, content filtering, or traffic engineering. These services may be either deployed at the CPE or, given expertise within the organization in question or at the edge of the provider's network. The latter option is more applicable for those organizations without the in-house expertise to manage the different services properly. This does not even take into account up-front costs for the discrete components, costs that may be amortized among many subscribers by deployment within the network.

Remembering that ATM is the encapsulation of choice across ADSL, nothing should preclude the support of nondata services at the CPE. In fact, given support for different ATM QoS on the DSLAMs and in the network core, an ATU-R could support native ATM services such as VTOA across AAL1 or AAL2, or MPEG-2 over AAL5. A settop box (described earlier and depicted as Scenario 5 in Figure 3.3) is in fact one variation. More common in those locations where multiservice ATM is widely deployed will be routers with analog or digital voice interfaces mapped into ATM.

Now considering the ATU-R's physical characteristics, the device must be available in a form factor suitable for the intended market. For example, an ADSL router for a branch office will have a very different size, power requirement, and cost than one for the residence. Some ADSL providers have in fact issued guidelines relating to these physical characteristics. Table 3.3 summarizes some of these possibilities.

TABLE 3.3. CPE PHYSICAL CHARACTERISTICS

Deployment	Number of Users	Power	Modularity	Current Cost
Residential	1–4	<10W	Not required	Sub-$500 Sub-$200 if PC NIC
Telecommuter	1–4	<10W	Not required	Sub-$500
Small Branch	max 10	27W	Some; VoIP-capable	Sub-$1500
Larger Branch	max 50	70–140W typical	Yes; VoIP-capable	$2000+

3.2.4 POTS Splitters

The last device through which the ADSL signal passes before leaving the customer premises on its way to the Central Office is the POTS/ISDN splitter (Figure 3.2). When the
copper pair enters the home or business, it contains both the ADSL data and the baseband
telephone signal, carried in different frequency bands. The splitter divides these two bands,
separating the POTS or ISDN traffic from the ADSL data through the use of filters.
Figure 3.4 illustrates the frequency spectrums of the data and voice signals.

FIGURE **3.4.**

POTS and ISDN
Spectrums

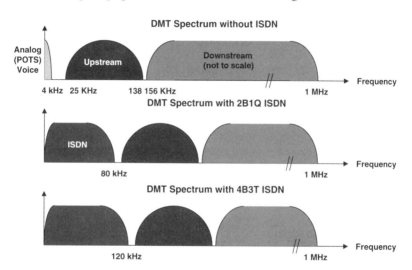

The POTS splitter contains a low-pass filter (LPF) that allows the voice frequencies to pass
to the phoneset, while filtering out the higher ADSL data frequencies. Within the ATU-R,
you will deploy a high-pass filter (HPF), this time eliminating the lower frequencies. Note
that some ATU-Rs integrate both the high-pass and low-pass splitter functionality, while
UADSL goes a long way toward eliminating the need for traditional filters. Here, small in-
line microfilters may be installed in parallel with the phoneset and the phone jack if
required. Although the focus here is on POTS splitters, in some countries ADSL must co-
exist with residential ISDN service. The splitter design in this case is a bit different and
more complex because the ISDN signal occupies a larger portion of the frequency spectrum
and the filter must be capable of attenuating the voice signal in a smaller frequency range as
illustrated in Figure 3.4.

The design and operation of POTS splitters is probably one of the least understood areas of
DSL deployment due to the background of most internetworking personnel. As discussed, a
splitter may contain an LPF, an HPF, and an overvoltage protection stage (see Figure 3.5).

FIGURE 3.5.

POTS Splitter Design

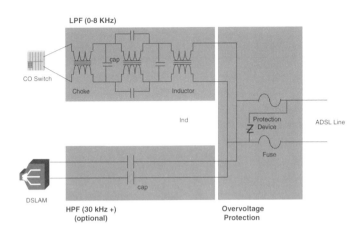

As Figure 3.5 illustrates, the LPF connecting to the CO Switch (or phoneset in the case of CPE) is constructed with a combination of chokes, inductors, and capacitors, whose net result is to filter all higher frequency traffic. In the case of POTS, a filter with an 8 kHz cutoff based on a fifth order design is acceptable. In many European countries, a 12 or even 14 kHz billing tone is also present, raising the bar on POTS splitters still further. However, the real complexity is when the provider attempts to run ADSL in the presence of ISDN. As described in Chapter 2, the entire DMT spectrum is shifted upward to allow for the 80 or 120 kHz ISDN spectrum. This of course requires deployment of a splitter operating in this frequency range.

The splitter will also include an optional HPF, which connects to the DSLAM (or PC), more applicable in the Competitive Local Exchange Carrier (CLEC) environment where loop testing when there is a fault may be more difficult. The HPF is just a set of capacitors in parallel with the circuit. Finally, an overvoltage protection circuit protects the splitter and ADSL modems from any voltage spikes on the loop. The most efficient design is to use a specific protection circuit bridged between the Tip and Ring of the local loop. This is in addition to the in-line fuses on the copper pairs.

In the United States, the POTS splitter will be installed between a Network Interface Device (NID), the demarcation point between the provider's physical network and the customer premises, and the ATU-R.

The splitter may contain both an HPF and an LPF or alternatively, only an LPF, allowing the POTS frequencies of 0–4 kHz to pass to the phoneset. In the latter case, the complete frequency spectrum across the copper will pass to the ATU-R, which must implement the HPF.

Although the splitter may implement the HPF, the ADSL Forum recommends that all ATU-R vendors implement this functionality. These options enable you to deploy the splitter and ATU-R in a number of configurations as illustrated in Figures 3.6–3.10.

The most basic configuration (Figure 3.6) involves installing a standalone HPF/LPF filter, connecting to both the in-house wiring and the ATU-R. This allows the ATU-R to be located near the TE, and the splitter may be preprovisioned. However, if the splitter is active, the user must provide power, and there is no guarantee that one vendor's HPF implementation won't adversely affect the ATU-R from another. This scenario also depicts the type of phone jack one would expect within the residence.

FIGURE 3.6.

Splitter Deployment Scenario 1

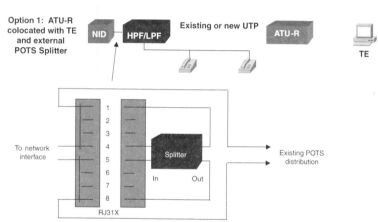

Note: Upon insertion of splitter plug, shorts from 1–4 and 5–8 are removed

If the HPF is located within the ATU-R, this possible problem is eliminated (Figure 3.7). This configuration will probably be most common.

FIGURE 3.7.

Splitter Deployment Scenario 2

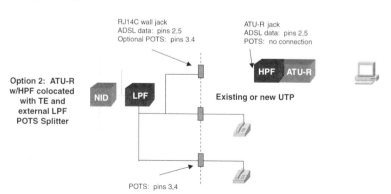

An alternative is to equip the phonesets with in-line LPFs, still locating the HPF at the ATU-R (Figure 3.8). This eliminates the central splitter but introduces additional complexities. For example, the frequency characteristics of the in-house network or improper LPF installation may negatively impact ATU-R performance. For this reason, this topology is not recommended.

FIGURE **3.8.**

Splitter Deployment Scenario 3

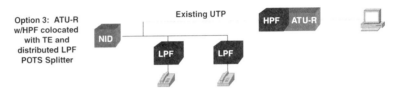

Option 3: ATU-R w/HPF colocated with TE and distributed LPF POTS Splitter

Yet another solution is to combine the ATU-R with the filters, allowing one to connect the in-house copper to a port on the ATU-R (Figure 3.9). However, this requires installation of the ATU-R at the copper ingress point where power may not be available or which may be environmentally unfriendly. In addition, the local interface from the ATU-R now requires extension to the TE.

FIGURE **3.9.**

Splitter Deployment Scenario 4

Option 4: ATU-R with integral POTS splitter colocated with NID

Finally, an option is to install new in-house wiring from the NID to a remote ATU-R implementing both the LPF/HPF, then connecting this to the existing copper (Figure 3.10).

FIGURE **3.10.**

Splitter Deployment Scenario 5

Option 5: ATU-R with integral POTS splitter colocated with TE

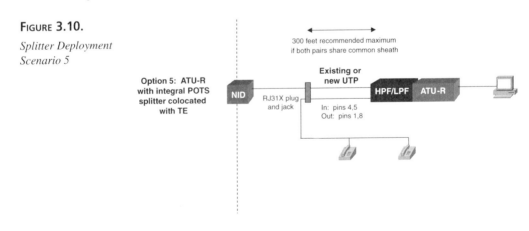

This new installation should be based on UTP Category 5, the same type of cabling used with Ethernet. Complexities here include ensuring that the ATU-R may not be disconnected and ensuring that there is no crosstalk between the incoming and outgoing pairs from the ATU-R. For example, if both pairs are in the same physical cable, the POTS ringing, dialing, and hook switch signaling noises are cross-coupled with the low-level ADSL receiving signals. In addition, this architecture is not suitable for a provider-owned and powered active splitter. Note that if POTS is not a requirement, the solution is much simpler, with the ATU-R always located near the TE.

Within Europe, the deployment possibilities are a bit more limited. In most cases, the POTS/ISDN splitter will be colocated with the ADSL NT, both devices owned and operated by the ADSL provider. The ATU-R then connects to the PDN, if there is one, while the phone traffic passes over the existing in-building cabling. Therefore, Figure 3.6 is most relevant.

3.3 Central Office Equipment

Upon leaving the customer premises, the first item encountered at the central office is the Main Distribution Frame (MDF) (referencing Figure 3.1 once again) where baseband POTS/ISDN traffic is split from the ADSL data. This voice traffic is carried to the existing circuit switch, while the data enters the ADSL modem rack, known as a Digital Subscriber Line Access Multiplexer (DSLAM). Note that at some point in the future, the DSLAM may be capable of intelligently handling the traditional voice traffic, consequently presenting a TR-303 or V5 trunk to the circuit switch. As opposed to the discrete ADSL modems on the user side, a typical DSLAM supports upwards of 64 modems in a chassis. This DSLAM multiplexes the user data onto an ATM trunk, presenting it to a local aggregation function, or alternatively, to an upstream aggregator reachable across an ATM access network (as Figure 3.1 illustrates). Depending on the service offering, the aggregator may provide access to local content and caches or simply forward the traffic at the ATM layer or Layer 3 across an upstream core network.

3.3.1 MDFs and Splitters

As with the customer POTS or ISDN splitter, the provider must also separate voice traffic from the data within the Central Office. At the CO, the copper bundles (each containing 50–1000 or more pairs) terminate on the Main Distribution Frame (MDF). This rack acts as a copper cross-connect between the local loop and the in-building plant, allowing for repair, troubleshooting, and user provisioning. The provider now has two options for ADSL subscribers. The first, and more optimal solution, includes deploying splitters fitting the profile of the MDF, allowing the split to occur within the single frame. If this is not possible, pairs assigned to ADSL subscribers extend to a splitter colocated with the DSLAM. The voice traffic then returns to the MDF over additional pairs, where it is then sent to the circuit switched network. This option introduces additional cabling complexity. Note that a third option is to dispense with splitters entirely, deploying ADSL as a second-line service or offering VoIP as a tariffed service (and described in Chapter 4).

The actual splitter may be provided by the DSL equipment vendor, although it is expected that a number of providers will have standardized on specific splitter solutions, sometimes in conjunction with the MDF. Some relevant standards in the splitter space include ADSL 97-189, T1-413 Issue 2, and GR1089-CORE. As introduced in the previous paragraph, next generation DLSAMs are expected to integrate intelligent voice processing, allowing the service provider to dispense with the splitter rack. Here, the DSLAM will present a T1 or E1 signaled interface to the existing PSTN switch. This would of course imply that the DSLAM has the hardware and software resiliancy to provide lifeline phone services since it will now be in the primary voice path.

3.3.2 DSLAM (ATU-C or ATU-LT)

To date the DSLAM has received the greatest amount of interest and coverage within the industry due to the crucial role it plays within an ADSL deployment. This is also the most crowded segment within the industry, with a number of vendors, both large and small, competing for a piece of what is expected to become a very large market in the coming years. After passing through the MDF and the POTS splitter (where installed), the copper pairs terminate at the DSLAM, and in most cases, each pair terminates at an individual modem. This statement is qualified with "most," since at least one DSLAM implements a modem-sharing architecture whereby the copper pairs first pass through a switching stage with users assigned to any available modem. This allows the provider to implement oversubscription, much like that which exists within Internet dial-up, and is described in section 3.3.2.4.

Within a typical DSLAM (Figure 3.7), an ATM backplane connects the CAP or DMT ADSL modem cards, a primary and optionally redundant control processor (1), and a primary and optionally redundant ATM trunk module. Data enters the DSLAM via connections to the POTS splitters (2) and is then routed to each of the modem cards (3). These perform the necessary processing, as illustrated in Figure 3.7 and described in the text that follows.

The data passes across the DSLAM backplane (4) and into the ATM trunk module (5). This module connects to either a colocated aggregator (described in section 3.4, "Aggregation") or an ATM access network. Note that there is a differentiation between ATM access and core networks, as depicted in Figure 3.11. An access network is located between the terminating

FIGURE **3.11.**

DSLAM Architecture

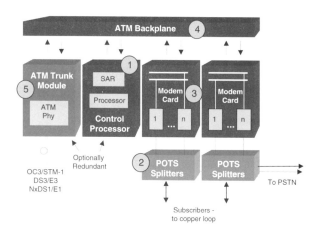

Central Office and a PoP, providing Layer 2 transport, while a core network is upstream from a Layer 3 service point. If the DSLAM implements signaling, signaling traffic (that is, VCI=5) will pass to the control processor. In this case, the control processor will play a much larger role in establishing the dataflow between the ADSL modem cards and the trunk. Note that even in DSLAMs that implement signaling, user traffic will not pass directly from one ADSL modem to another—it must first pass through an aggregator or ATM switch via the trunk module.

The ATM trunk module may be OC-3/STM-1, DS3, E3, or even NxDS1/E1 in smaller locations or DLC environments. Table 3.4 characterizes a typical DSLAM OC-3/STM-1 uplink.

TABLE 3.4. DSLAM UPLINK CHARACTERISTICS

		Minimum	*Typical*	*Maximum*	
	Wavelength	1260	1310	1360	nm
Tx	Multimode	−20	−17	−14	dBm
	Singlemode	−15	−11.5	−8	dBm
Rx	Multimode	−30		−14	dBm
	Singlemode	−28		−8	dBm

The amount of buffering within the modem and trunk cards on a DSLAM is also important if it is to support bursty LAN data or real-time traffic. Most first generation DSLAMs supported only a single QoS, UBR, frequently presenting a problem to the user's traffic if equipped with little or no buffering. One of the better equipped DSLAMs during this timeframe offered adequate buffering per modem in the downstream and upstream direction to handle short data bursts. Later, DSLAMs appeared capable of supporting multiple QoSs.

3.3.2.1 DSLAM Standard Compliance

Since the DSLAM is deployed within the transmission area of a Central Office, it must comply with stringent telco/PTT requirements in areas such as packaging, heat dissipation, power, and safety. Every DSLAM designed for the CO conforms to a 12 inch/300 mm depth and either a 19 inch (United States and the rest of the world) or 23 inch (United States) width, allowing deployment within the transmission frame room. The DSLAM operates from 48 VDC station power and is designed to dissipate no more than 530 watts within a 7 foot rack. Typical operating temperatures range from 5 to 40°C (operating), −5 to 50°C (short-term operating) in a 5–90 percent noncondensing environment. Note that this contrasts with DLC-mounted DSLAMs complying with TR-NWT-057 for outside plant. Here, −40–65°C (operating) is the norm. Figure 3.12 depicts a typical central office DSLAM, comparing it with one more suitable for DLC deployments in the lower half of the figure and described in Section 3.6, "Digital Loop Carrier Systems and the Full Service Access Network."

FIGURE **3.12.**

DSLAM Dimensions

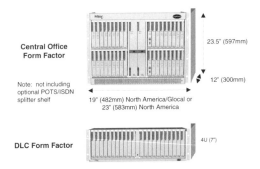

Central Office
Form Factor

23.5" (597mm)

Note: not including
optional POTS/ISDN
splitter shelf

12" (300mm)

19" (482mm) North America/Glocal or
23" (583mm) North America

DLC Form Factor

4U (7")

In the United States, the relevant standards are known as Network Equipment Building Standards (NEBS), the most stringent being Level 3, while Europe looks to the European Telecommunications Standardization Institute (ETSI) for guidance. The most relevant standards include UL 1950, UL 1459, FCC Part 15, Bellcore GR-1089-CORE, and GR-63-CORE, detailed in the appendix as they relate to NEBS compliance. Although not a requirement per se, almost every DSLAM is equipped with redundant power entry modules, and most replicate control processors and ATM uplinks.

3.3.2.2 DSLAM and ADSL Modem Concerns

Within a Central Office environment, power consumption is a major issue, with equipment and racks limited in both wattage and heat dissipation. In early ADSL deployments, this presented a bit of a problem due to the high power requirements of the ADSL modem chipsets, resulting in lower acceptable density. For example, first generation CAP chipsets required 3 watts at full BW, while DMT drew even more at 3.5 watts for the ADSL transceiver and analog front end. Luckily, these figures have improved dramatically over the last year or two, with the latest DMT (Issue 2) chipsets now requiring just 2 watts per user. This has allowed densities to increase from the order of 240 per 7 foot rack to 480 or even 720 modems. In addition, as DSPs continue to evolve, further optimizations are possible. For example, a single DSP performing the ADSL processing could drive multiple analog front ends. An example of this is the Texas Instruments TNETD3000® chipset.

The actual ADSL modem set integrates a number of components, depicted in Figure 3.13.

FIGURE 3.13.

ADSL Modem
Components

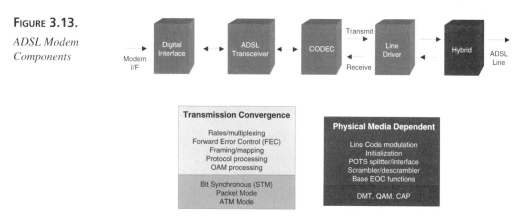

Source: ADSL Forum

Traffic arriving into the DSLAM across the ATM trunk first passes through the demultiplexing function on the backplane and then into the digital interface on the modem. This provides the necessary bit stuffing/robbing, elastic store, and digital processing (CRC detection, scrambling, interleaving, and FEC) before passing the data to the ADSL transceiver stage. This device, based on custom hardware or a DSP, actually encodes the data and performs the Inverse FFT (Fast Fourier Transform) if part of a DMT-based system. Note that this stage does not exist within CAP/QAM. It then forwards the signal to the CODEC, responsible for performing Digital to Analog Conversion (DAC) and filtering before passing the data to the line driver. This unit actually powers the ADSL line and consumes the greatest amount of power in comparison to the other components. Finally, the signal reaches the hybrid circuit that includes the two wire loop interface. The lower part of Figure 3.9 depicts the functions performed at both the transmission convergence (TC) and physical media dependent (PMD) layers within the modem, TC mapping to the digital interface and PMD mapping to the remaining elements.

3.3.2.3 Subtending

An alternative to connecting every DSLAM to the aggregator or the ATM access network is to implement a technique known as subtending. Here, the uplinks from some DSLAMs, either within the same CO or in remote COs, connect to other DSLAMs. This may be useful where some low-density COs have only a single DSLAM. As an example, remote DSLAMs may subtend via DS3 or even multiple DS1 (ATM IMA) uplinks, with only the furthest upstream DSLAM presenting an OC-3 into the core or to the aggregator (see Figure 3.14).

FIGURE **3.14.**

Subtending

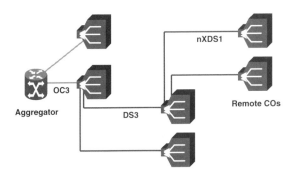

Alternatively, subtending within a CO may be useful in conserving ATM switch ports. Here, a rack (or more) of DSLAMs may present a single uplink to the aggregator. Subtending is also expected to be of major importance within ADSL deployments based on Digital Loop Carrier (DLC) systems due to the large number of remote DSLAMs deployed serving small subscriber groups. Although subtending was not widely supported during 1998, it should become more common during 1999.

3.3.2.4 Modem Sharing and DOH

Revisiting modem sharing, due to the high cost of ADSL chipsets, service providers attempt to mimimize the cost of service deployment. One way to accomplish this is by allowing a single modem to be shared among multiple subscribers by deploying a switching stage between the incoming copper pairs and the modems. For example, a provider may deploy a DSLAM with 32 modems serving a total of 64 subscribers, resulting in a 2:1 oversubscription. The trick here is to implement sharing while still providing the "always-on" service expected of ADSL subscribers. An approach in optimizing user satisfaction is to implement tiered tariffing, whereby premium subscribers may have dedicated modem access, while the provider may push oversubscription even to 5:1 for the most basic of subscribers. For comparison, the average Internet dial-up service implements up to 10:1 oversubscription, while corporate modem pools implement substantially lower oversubscription.

Figure 3.15 depicts the dataflow within this type of DSLAM.

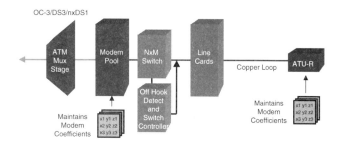

FIGURE 3.15.

Modem Sharing

The copper loops terminate on concentration modules, which then connect to a switching fabric. When a user activates the CPE, the fabric connects to an available ADSL modem, forming the end-to-end path. An important consideration in implementing this technology is modem retraining time. Traditionally, when a user initiates a connection to an ADSL modem for the first time, the activation sequence could take up to 30 seconds as the modems exchange coefficients and negotiate the line rate. The success of DOH is dependent upon storage of the necessary parameters between the CPE and the CO modems, allowing the modems to retrain in the order of seconds.

3.3.2.5 DSLAM in PON/FSAN Systems

Another area of interest is compatibility of DSLAMs with Digital Loop Carrier (DLC) or Passive Optical Network (PON)/Full Service Access Network (FSAN) systems. The former terminology is used within North America, while the latter is more common in the rest of the world. Although most DSLAM installations are expected to take place within the CO, nothing precludes the deployment of smaller, hardened DSLAMs within pedestals closer to the end users. Here, the copper extends much less than a mile, with the aggregate back-hauled to the CO via fiber or multiple T1/E1 links. This allows the provider to extend the reach of ADSL considerably compared to a CO-based solution. The coexistence of ADSL and DLC or PON/FSAN is actually of great importance as more providers deploy this technology as part of their copper plant upgrades. In contrast to CO DSLAMs, which must support upwards of hundreds of subscribers within a rack, the typical remote system may serve on the order of 16 or 32 subscribers. Note that a DSLAM located within a pedestal will have different physical requirements than one deployed in the CO. Three of the most notable parameters are temperature operating range, power consumption per modem, and, closely related, heat dissipation. ADSL technology could even move out to the pedestal serving a very small set of subscribers given solutions to these requirements, although VDSL, described in Chapter 6, could be a better fit.

3.3.2.6 Transmission Speed

As important as modem footprint and power consumption is transmission speed, a function of the modem technology (including power), loop length and gauge, and the presence of ISDN splitters. ADSL relies on a technique known as rate adaptation whereby the modems at either end of the loop negotiate the maximum achievable bandwidth. This will of course vary on a loop-by-loop basis, but even across a given pair, the maximum rate may change, depending upon external influences such as the weather. This presents some interesting problems for the ATM layer, described in the next chapter. The service provider will in most cases set the maximum transmission speed to a value less than that obtainable across the loop, also basing this setting on the user's service profile.

Table 3.5 lists predicted transmission speeds for CAP across different loop lengths, and Table 3.6 lists these speeds for DMT. Note that these are only estimates as the actual rates achieved vary widely by installation and even by vendor implementation.

TABLE 3.5.   ADSL BANDWIDTH FOR CAP AND DMT

	Distance	Gauge/Diameter	Downstream	Upstream
CAP	9 Kft/2.7 km (CSA 6)	26 AWG/.4 mm	7.616 Mbps	680 kbps
	12 Kft/3.7 km (T.601–13)	24/26 AWG	2.040 Mbps	128 kbps

TABLE 3.6.   ADSL BANDWIDTH FOR DMT

	Distance	Gauge/Diameter	Downstream	Upstream
DMT	9 Kft/2.7 km (CSA 6)	26 AWG/.4 mm	6.144 Mbps	640 kbps
	12 Kft/3.7 km (T.601–13)	24/26 AWG	1.696 Mbps	160 kbps
	12 Kft/3.7 km (CSA 8)	24 AWG/.5 mm	6.144 Mbps	640 kbps

As vendors introduce more refined modems and as the service providers gain experience in scaling their loop qualification techniques and update their cable records (some in quite dismal shape), results will be more predictable. Please refer to the more detailed technical descriptions of CAP and DMT modems in Chapter 2, which looks at power, bandwidth, latency, and performance in the presence of disturbers.

The distances listed in Tables 3.5 and 3.6 correspond to a number of standard copper loop lengths as depicted in the table. Designators include CSA, for Carrier Serving Area and ANSI. A CSA0 test is effectively back-to-back, with a maximum of 10 feet. Note that even the maximum bandwidths are less than the 7–9 Mbps figures quoted in marketing literature, rarely achievable over actual copper plant.

Given the values in Tables 3.5 and 3.6, it is useful to look at the effect of ISDN splitters on the maximum bandwidth obtainable. Referencing Figure 3.4, note the frequency spectrum of the ISDN signal in comparison to POTS. This requires an upward shift in both the downstream and upstream ADSL data components, limiting the distance. Table 3.7 depicts the new values.

TABLE 3.7. ADSL BANDWIDTH AND DISTANCE WITH ISDN SPLITTERS

Bandwidth (Downstream/Upstream)	POTS	2B1Q ISDN	4B3T ISDN (Germany)
6 Mbps/640 kbps	3 km	2.7 km	2.5 km
2 Mbps/192 kbps	4.5 km	4 km	3.5 km

Note that alternative means of integrating the ISDN signal do exist, including compressing the ADSL frequency spectrum at the expense of bandwidth. This is less preferable to shifting the spectrum and, in fact, in most countries where ISDN is a major factor, the loop lengths between the CO and the user are shorter than in the United States.

You now have all the necessary components in-hand for deployment of the DSLAM within the CO: the actual modems, the splitter, and where required, the digital off-hook hardware. Figure 3.16 depicts the progression of dataflow across a typical 7 foot rack within a CO.

FIGURE 3.16.

Central Office DSLAM Deployment

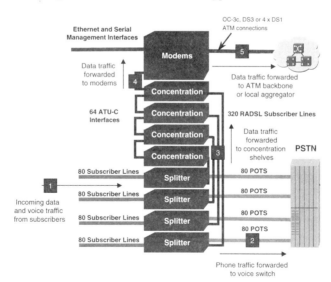

The dataflow pathway illustrated in Figure 3.16 is as follows:

1. Incoming data and voice traffic from subscribers flow across the copper loop and through the splitters.
2. Voice (PSTN) traffic is forwarded to the voice switch.
3. Data traffic is forwarded to concentration shelves.
4. Data traffic is forwarded to modems.
5. Data traffic is forwarded to the ATM backbone or local aggregator.

3.4 Aggregation

Although the ADSL CPE and DSLAM are the two most visible components within an ADSL deployment, they are only part of an end-to-end service architecture. This section and the sections that follow describe the additional components that are critical in allowing the ADSL provider to offer a successful service, or that permit an ISP to participate in such an offering. The first of these value-added systems aggregates the virtual circuits from the DSLAMs. If colocated with the DSLAMs, sometimes in the same racks, this system will be owned and operated by the ADSL provider. In the United States this system would be the ILEC, or a CLEC with colocation rights (described in section 3.10, "Regulatory Considerations"). In other countries this system would be owned by the primary PTT/telco. An ATM access network (see Section 3.5, "Access and Core Networks") or a DLC/FSAN system could also separate the DSLAM from the aggregator, although in this case it would still be operated by the ADSL provider. Alternatively, the aggregator could be operated by an ISP, terminating the ADSL provider's virtual circuits. Here, a given ADSL offering could be surrounded by aggregators owned and operated by a number of ISPs, each possibly providing services for other upstream ISPs as well. Figure 3.17 depicts deployment scenarios where the aggregator is deployed both within the Central Office and within the CLEC/ISP PoP.

FIGURE **3.17.**

Aggregator Deployment Options

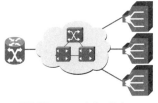
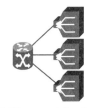

DSLAMs connect directly to
ATM access network

DSLAMs connect directly
to aggregator

The aggregator performs a new function, initially found within xDSL deployments but not limited to this technology by any means. The aggregator aggregates the ATM uplinks from multiple DSLAMs, passing the traffic into an upstream ATM or packet network after performing a combination of Layer 2 and Layer 3 services depending on the service offering. The aggregator therefore carries out two vital functions: protection of and optimization of expensive core bandwidth; and more important, allowing the provider to deploy efficiently the Layer 2 and Layer 3 services required by the different customer bases. The system should therefore include both ATM switching as well as routing capabilities, in effect replacing some of the discrete components currently found at the edge of a service provider's network. Interfaces on the aggregator in the direction of the DSLAM are usually ATM, though in at least one instance the access network converts the DSLAM's ATM trunks to Frame Relay before presenting them to the aggregator (this time operated by an ISP). In the upstream direction, interfaces may include ATM, Ethernet, Frame, PPP over SONET/SDH, and PPP over optical transport (bypassing the SONET/SDH layer).

Functions performed by the aggregator may include ATM switching, PPP termination into routing, PPP tunneling into L2TP/L2F/PPTP, bridging, and additional value-add such as firewalling, VPNs, and traffic management. When performing ATM switching, the device is expected to implement the necessary traffic management (TM 4.0 and updates), VCC support (PVCs, SVCs, soft-PVCs), and routing (PNNI). Termination of the user's PPP or bridging sessions allows the ADSL provider to offer access to local content where permitted. If the aggregator is operated by an ISP, this capability is, of course, a given. Tunneling provides secure access from the user to ISP or corporate gateways, analogous to currently deployed Internet dial VPNs. Here the system acts as an L2TP Access Concentrator (LAC) and may be deployed by the ILEC. A different service model offering some of the same functionality is known as PPP Termination Aggregation (PTA), mapping user PPP or bridged sessions into upstream routed connections (without L2TP tunnels). Finally, until the widespread deployment of PPP-capable CPE, bridging will remain a factor in ADSL deployments. If bridging traffic end-to-end or terminating the bridged traffic into routing locally, the system acts as a Broadband Access Server (BAS). Chapter 4 describes these functions in great detail and how they enable value-added services.

As with the DSLAM, the aggregator should comply with the relevant CO or PoP standards for a given region. For example, within the United States, compliance with NEBS Level 3 will allow deployment within an ILECs CO (also applicable to some colocated CLECs). Although requirements are not as stringent within ISP PoPs, a single hardware platform should be optimized in such a way that it is suitable for deployment in both environments.

3.5 Access and Core Networks

Leaving the Central Office environment leads to the provider's MAN or WAN core network. This infrastructure, based on ATM, packet, or in some cases, Frame Relay, provides connectivity to other ADSL providers, ISPs, and corporations. As opposed to an access network, which provides connectivity internal to the ADSL provider, a core network peers with external entities. The choice of core technology will depend on the type of peering required and the bandwidth required by the Internet offering. For example, one Regional Bell Operating Company (RBOC) may have only ATM connectivity to another, with the trunks used for both Internet and circuit switched traffic. Or, an ADSL provider may offer a very basic service, extending the user's VCCs across a service boundary to a second provider. Over time, as more providers migrate to packet-based cores for Internet traffic, peering is expected to occur over dedicated IP trunks. This is especially true at bandwidths in excess of 155 Mbps.

3.5.1 ATM Access and Core Networks

Since DSLAMs present an ATM uplink, this will be the technology of choice for ADSL access networks, connecting remote DSLAM to the provider's upstream CO, or alternatively, backhauling the VCCs to a PoP or to another provider. This use of ATM also helps the providers leverage their investments in ATM, since the same infrastructure will support Circuit Emulation Service, Frame Relay, in some cases narrowband voice, and now Internet connectivity. Although some DSLAMs are equipped with Frame uplinks, they are not expected to play a major role in public ADSL rollouts. Once the ADSL VCCs reach the PoP, they may extend across a second ATM network, this one operated by an ISP or CLEC. In fact, this architecture (as illustrated in Figure 3.18) may become common in the future. Chapter 2, describes ATM as a technology in greater detail.

FIGURE 3.18.

*ADSL and ATM
Deployment Scenario*

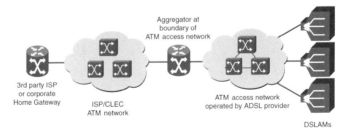

3.5.2 Packet-Based Core and Access Networks

A router core provides an alternative to the traditional ATM networks currently deployed. Although ATM will remain the technology used across the ADSL loop, as IP bandwidth needs increase, there is a growing move to transition IP trunks from the existing ATM backbones to a dedicated router infrastructure. In most cases, the inflection point is at about 155 Mbps, the lowest available bandwidth for PPP over SONET/SDH interfaces. At speeds in excess of 155 Mbps, and especially when Dense Wave Division Multiplexing (DWDM) enters the picture, dedicated trunking is the preferred solution. Note that this transition applies to the backbone. There will still be ISP and corporate Home Gateways (HGWs) connected to the provider's ADSL network at DS1/E1, nxDS1/E1, or DS3/E3. Here, ATM still makes perfect sense. Initially, the ADSL traffic will transition from ATM to packet at the PoP. However, as bandwidth needs increase, and the ILECs begin to deploy aggregation functionality within their access networks, the transition from ATM to packet could occur at a CO containing both a DSLAM and aggregator. Here, the DSLAMs present an ATM uplink to the aggregator, which then connects to a packet core via a Fast or Gigabit Ethernet, or even more so in the future, optical. Figure 3.19 depicts this transition.

3

FIGURE **3.19.**

Transition from ATM to Packet in the Access Network

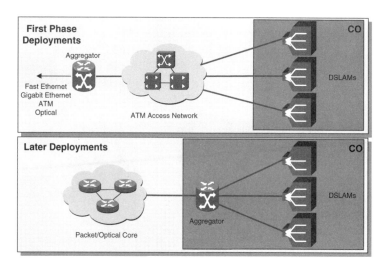

3.6 Digital Loop Carrier Systems and the Full Service Access Network

The various components of the end-to-end architecture are also closely associated with fiber in the loop deployments. Although this may at first glance seem to be at odds with ADSL, which most think of as a Central Office to CPE copper-based technology, in fact there is a great deal of compatibility between ADSL and Digital Loop Carrier (DLC) systems. In fact, these systems allow service providers to extend the reach of ADSL beyond the capabilities of some parts of the existing cable plant. In North America, the term DLC is usually used to refer to these systems which extend fiber from the CO to a point serving anywhere from eight to a few hundred subscribers. This fiber termination point provides access to voice, data, and even cable TV, multiplexing the signals via TDM or ATM technology onto an SONET/SDH uplink. Note that even with eight subscribers, this is not Fiber-to-the-Home (FTTH) setup; it still relies on the reuse of the existing copper into the home.

This architecture is known as the Full Service Access Network (FSAN), proposed by a group of service providers throughout the world. This introduces the concept of a Service Node (SN) within the network core. This Service Node is assumed to be an ATM switch, and in the direction of the subscriber, first connects to the Optical Line Termination (OLT), which in turn connects to an Optical Network Unit (ONU) via an Optical Distribution Network (ODN). The DSLAM, as it is known in the United States, is associated with the ONU, while the ODN could connect an ATM core network with each of the central offices. This architecture allows the ONU to be placed within the CO, or alternatively, closer to the subscriber, analogous to the DLC system. The copper pairs of course terminate at the ONU. Figure 3.20 illustrates an FSAN deployment.

FIGURE 3.20.

Full Service Access Network (FSAN) Deployment

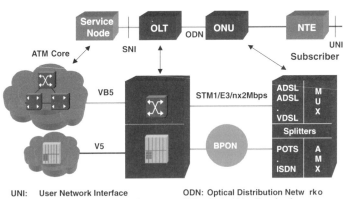

UNI: User Network Interface
NTE: Network Termination Equipment
ONU: Optical Network Unit
BPON: Broadband Passive Optical Network
MUX: ATM Multiplexer
AMX: Access Multiplexer

ODN: Optical Distribution Netw rko
OLT: Optical Line Termination
SNI: Service Node Interface

In an actual deployment such as that illustrated in Figure 3.20, the provider may first deploy an ATM core connecting directly to the ONU (DSLAM). Splitters at the DSLAM allow voice connectivity as described earlier, while the subscriber's NTE connects to the ONU across the local loop. Later, the provider deploys an OLT, connecting to both the ATM core via SDH and the narrowband circuit switched core. This OLT integrates ATM switching and voice termination functionality. SDH, acting as the ODN, connects this OLT with the ONU, which now integrates both voice and data termination capabilities. This ONU could be located much closer to the subscriber. The final phase of deployment sees the replacement of much of the narrowband network by functionality within ATM. The OLT becomes a full ATM switch, connecting to the core network via the VB5 interface, while within the fiber loop we introduce the concept of a Broadband Passive Optical Network (BPON), interconnecting the OLT with the ONU. Service providers are expected to deploy this last phase after the year 2000.

Standardization activities within ETSI supporting FSAN include groups focusing on access transmission systems (STC TM6), architecture and functional requirements of broadband access networks and functional characteristics of user interfaces (STC TM3), and VB5 interfaces (STC SPS 3) which are described in Chapter 4. Note that this loop architecture also lays the groundwork for the eventual deployment of VDSL, described in Chapter 5.

3.7 ISP/Corporate Gateways

After traversing the core network, the user's data finally arrives at the ISP or corporate gateway. This device terminates any ATM VCCs, PPP sessions, or L2TP/L2F tunnels, passing the traffic into the provider's core or corporation's intranet. User authentication commonly occurs at this boundary, described in the section on provisioning and within Chapter 4 on end-to-end services. The actual hardware device performing this session termination function is commonly known as a Home Gateway (HGW), or in the L2TP tunneling space, an L2TP Network Server (LNS). When terminating PPP, this device must have the scalability to handle a large number of user PPP sessions. Unlike the customer, connected to the network via ADSL, the HGW will connect natively to an ATM or packet core. As an example, in early ADSL rollouts, the ISP, or corporation, leased a DS1, DS3, or OC-3 ATM connection to the ADSL provider. This required HGW scalability, as an ISP with traffic to justify only a DS1 connection would not want to be forced into deploying a box sized for OC-3. HGW offerings may therefore be judged on their uplink bandwidth, uplink type, ISP/corporate facing interfaces, session scalability, and packets-per-second (pps) performance. The HGW will usually be managed by the ISP or corporation, although the ADSL provider may have some role in monitoring as well.

3.8 Content, Caching, and Gateways

Although many focus on the "physical" infrastructure of ADSL, there are other components critical to a service-rich offering. Content servers, operated by the ADSL provider or by a third party, allow community of interest information to be injected into the network. The most logical place for these servers is as close to the user as possible at the edge of the network, conserving core bandwidth. Types of content servers may include video streaming (both IP multicast and ATM) as well as push servers aimed at specific market segments such as the financial community.

A second vital component is the web cache. This device allows the provider to cache web traffic generated at distant sites, conserving core bandwidth, and minimizing the server response time from the user's perspective. The web cache helps to make ADSL "feel" like a premium service.

Closely related to web caches, though operated by corporations or ISPs, are web traffic redirectors. As an example, a global corporation may establish a web presence in a number of countries. Users are then pointed to the nearest server based on their location, minimizing response time and helping to distribute server load.

Finally, gateways allow service interworking between different technologies. The most well known of these is probably VoIP gateways, permitting the service provider to offer a transparent end-to-end voice service spanning both the Internet and the classical circuit-switched voice network. VoIP gateways may be deployed by the ADSL provider as part of a premium service offering, by an upstream ISP, or by the corporation.

These various service engines add value to the ISP's ADSL offering and help enhance the business model. In fact, ISPs will probably need to differentiate themselves in this way if they are to survive in an increasingly competitive environment.

3.8.1 Web Caching

Consider a user accessing a distant web server across an Internet backbone. Even if the user is connected to the local ISP via a high-speed connection, the server response time will still depend on the distance to the server (in ms), end-to-end available bandwidth, a product of any congestion, and the loading on the server. Ideally, popular web pages could be cached locally, reducing the response time and offloading the Internet backbone. This is the concept behind web caches, which intercept HTTP requests and then forward them to the distant servers as illustrated in Figure 3.21.

FIGURE 3.21.

Cache Data Flow

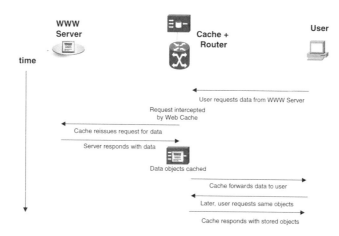

The returned data passes through the cache and, if "cachable," is retained for future use. The term *cachable* refers to objects that can be stored locally, since some data is tagged in such a way to preclude caching, either due to timeliness or copyrights. In any case, if even a small portion of the total web traffic is cached, the bandwidth, and therefore cost savings, may be substantial.

Figure 3.22 illustrates a hierarchy of cache servers, one set at the ISP's international boundary, and the other local to the ADSL service.

FIGURE 3.22.

Hierarchical Cache Architecture

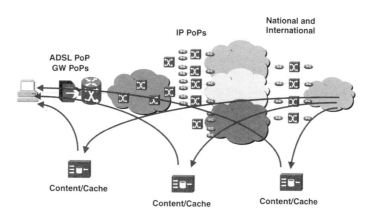

Remember that ADSL is a "premium" service. If the user experience is impacted by factors beyond the control of the ADSL provider, such as distant ISPs with slow backbones or overloaded caches, the success of the service will be called into question.

3.8.2 Video Servers

Video servers provide access to local multimedia content, either general in nature or targeted at a specific audience. Although these video servers appear to be deployed at the ADSL PoP in the architecture figure (Figure 3.1), they may of course be deployed by any of the upstream entities: the ISPs, corporations, governments, and educational institutions.

There are two major types of video servers. The first is IP-based, relying on IP unicasting or multicasting (preferred) for video distribution. Here, user access to the stored or live video stream is via a standard browser interface. If the stream is destined for a group of users (for example, a broadcast or rebroadcast of a corporate address), it will be associated with an IP multicast (IPmc) group. Alternatively, an individual user could request a specific program, the "video-on-demand" paradigm. Here the program would be unicast across the network. The second option obviously has a much greater impact upon bandwidth than the former, although the ADSL environment equips the PoP appropriately. Running high-quality unicast video across a WAN link may be another story, especially if sending multiple streams in parallel. IPmc is capable of handling broadcast-quality video, although this fact is not well known.

The second type of video server is ATM-based where the video traffic, usually MPEG-2 (broadcast quality), is encapsulated directly over ATM AAL5 and forwarded across the network. Just as with IP-based video, replication mechanisms within the network will optimize bandwidth. The ATM-based video server relies on ATM's point-to-multipoint capabilities, where the ATM switch performs the multicast replication. Note that this technique may also be used to optimize the distribution of IPmc across an ATM network given proper multicast software. In an ATM-based video server environment, each user participating in the server is provided with a dedicated VCC to the server. The user, through a web interface, for example (as the control traffic will almost always be carried as IP), will request a given broadcast which is then transmitted across the circuit.

Both IP and ATM-based video servers are expected to be widely deployed—the former used more for communities of interest, and the latter for general entertainment. Figure 3.23 illustrates both the ATM and IPmc-based video possibilities.

Note the differences in PVC utilization. In the case of IP multicasting, both the video and the nonvideo data traverse the same PVC, with QoS provided by Layer 3 mechanisms such as CAR, RSVP, and IP precedence. In contrast, when deploying ATM-based video, this traf-

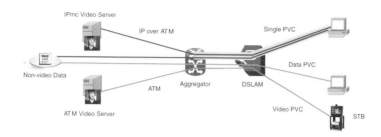

FIGURE 3.23.

*Video Service
Architecture*

fic will normally occupy a dedicated PVC with a different QoS than the data PVC. This scenario therefore has an impact on PVC provisioning.

3.8.3 Push Servers

The next type of local content server is the push server, at present most prevalent in the financial sector although this is not expected to remain the case in the future. Unlike standard web servers where the user must "pull" the information via a browser, a push server automatically delivers the data to the user based on a set of subscription parameters. For example, a stock trader may require a set of stock quotes and currency exchange rates to conduct business. Here, the push server will automatically (and on a timed basis) distribute the necessary data to the trader. One of the most widely deployed manifestations of this architecture is TIBCO's information bus, operating across both the MAN and WAN. Although the focus is initially on financial users, nothing precludes reuse of the same techniques in the dialup environment. In an ADSL deployment, the push data will follow much the same path as the multicast data, passing through the service aggregator, across the DSLAM, and to the subscriber. In fact, data destined for groups of subscribers will use IPmc and its reliability extensions for distribution.

3.8.4 VoIP Gateways

Last, but surely not least, the ADSL provider may deploy VoIP gateways as part of the service offering as shown in Figure 3.24.

Here, VoIP traffic generated by the user (1 in Figure 3.24) and carried across the ADSL loop is forwarded by the aggregator to a telephony gateway connecting to the PSTN or an ATM switch. This device performs the necessary voice encoding/decoding through the use of DSPs, as well as interpreting various voice signaling protocols. At the signaling layer, the gateway will provide interworking between Signaling System 7 (SS7) and either H.225/Q.931 for H.323 clients or alternatively, the Session Initiation Protocol (SIP) if based on

FIGURE **3.24.**

VoIP GW Architecture

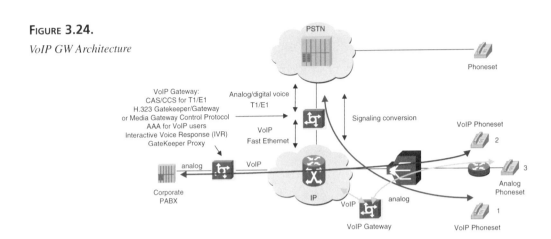

IETF protocols. In both cases, internal synchronization is the responsibility of the Media Gateway Control Protocol (MGCP). The control element also maintains the dialing plan, handles accounting and fault isolation, and ensures that the proper resources are available across the IP network to maintain voice quality. From this gateway, the analog or digital voice traffic passes into the existing circuit switched network.

As an alternative, a user's corporate phone service may be converted into VoIP, also at the CPE, for transport across the ADSL loop.(2 in Figure 3.24). This VoIP traffic would then be converted back to analog or digital voice at the corporate PABX. Such an architecture could be useful in the telecommuting environment, for example, where a worker accesses the corporate voice network via a VoIP connection across ADSL. Over time, more and more of the voice traffic may remain on the ADSL/Internet service as voice management and gateway systems grow in sophistication.

The reverse of these scenarios would be to transport the existing analog phone traffic within the baseband of the local loop via POTS splitters. At the CO, it would be converted into VoIP for transport across the Internet backbone (3 in Figure 3.24). By mid-1998, a number of telcos and ISPs had announced this service in the non-ADSL space.

Initially, the system supports CAS/CCS on the circuit switched side and participates in the H.323 infrastructure toward the Internet/intranet core. As vendors introduce additional functionality such as the integration of SS7, more advanced intelligent networking features will be available across the gateway. Ultimately, a user will enjoy the same features across VoIP as available across a traditional corporate PABX. In fact, VoIP phonesets that are almost identical in features and performance to the typical PABX extension exist today— the only difference is an Ethernet interface instead of the RJ11.

3.9 Management and Provisioning

Within an ADSL deployment, the responsible service providers and/or ISPs will be required to manage the end-to-end system at a number of different logical layers, frequently with different goals. These layers conform to the Telecommunications Management Network (TMN) model, described in section 3.9.1.

Moving up a layer, the providers require a way to manage the interconnections between these different network elements. Here, configuration and monitoring capabilities exist at different layers of the protocol stack, including the ADSL, ATM, IP, and session layers. This network management provides for an end-to-end view of the infrastructure. However, different organizations such as the ILECs, CLECs, and ISPs may have responsibilities for different components, complicating matters. As an example, one provider may manage the ATM switches and DSLAMs, while another may have responsibility for the CPE and routers. Across the ADSL link, G.997.1 and the associated ADSL Line Management Information Base (MIB), under development within the ITU-T and the ADSL Forum respectively, address the ADSL and ATM layers, while at present, vendor proprietary systems apply to the IP layer (Layer 3) and above.

Taking a more subscriber-centric view, the network manager must be able to configure effectively the various parameters associated with the users of the system. This leads into the requirement for flow-through provisioning, where single actions may impact the configuration of a number of discrete systems. At this layer, interfaces begin to appear into the provider's service and business management systems, while providing Customer Network Management (CNM) capabilities when required.

A parallel issue is the requirement to integrate any ADSL-specific management systems with existing architectures, or at least provide the necessary open interfaces. In the remainder of this section, these various existing requirements and details and planned solutions are described.

3.9.1 Telecommunications Management Network

The Telecommunications Management Network (TMN) (Figure 3.25) provides a layered management architecture which may be applied to ADSL.

At the lowest layer is the Element Layer (EL), relying on instrumentation of the Network Elements (NE) via MIBs. Functions here include configuration, troubleshooting, and performance monitoring. Currently, only the ADSL Line MIB is standardized, and any addi-

FIGURE **3.25.**

TMN Framework

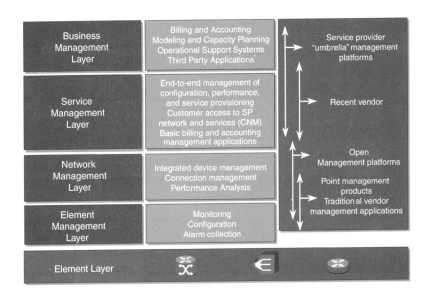

tional instrumentation relies on vendor-specific MIBs. A goal is to combine the relevant elements of these proprietary MIBs into a standard subnetwork MIB.

The next layers are the Element Management Layer (EML) and Network Management Layer (NML). The EML consists of the various vendor-supplied element management systems for groups of NEs. The NML is usually provided by the service provider, interfacing to the vendor's EMLs. The next layer of the hierarchy is the Service Management Layer (SML), which includes interfaces to customer databases and the provider's billing system. Finally, the Business Management Layer (BML) extends across the provider's entire organization, interfacing with 3rd parties where required. There is a proposal to combine the EML and NML functionality in the ADSL space, where the vendors would then provide the necessary functionality to allow the service provider to manage the ADSL network as a single entity. This enhanced NML would then interface with the provider's existing SML.

Parallel to the TMN hierarchy are the various functions performed by a management system. This is the classic "FCAPS" set of functionality: fault, configuration, accounting, performance, and security. For ADSL, configuration may consist of NE installation, image storage, ISP service provisioning, customer service provisioning, and preprovisioning of services and equipment. In the area of fault management, typical functions include service outage reporting, trap filtering, verification of end-to-end ADSL connectivity, fault localization, diagnostics, and automatic restoration. Finally, performance management may include setting thresholds, conducting assessments of network performance and then forwarding this data, or correlating performance data between the different protocol layers.

A set of ADSL Forum Network Operations Reference Model (NORM) documents, outlining top-level requirements, summarize these various functions Two under development include "Repair and Maintain Service" and "Provide Customer Service."

3.9.2 Management Protocols

At the lowest layer of the network management architecture, administrators have a requirement to configure, monitor, and troubleshoot individual network elements. Although many available management platforms provide only this functionality, as described in the previous section, it is only one part of a total management solution. In order to manage a switch, router, or modem effectively, a management protocol must include some form of instrumentation allowing it to communicate with a network management platform. The most common protocols are the Simple Network Management Protocol (SNMP), the Common Management Information Protocol (CMIP), and newer web-based techniques. More recently, a simplification known as the Simple Device Protocol (SDP) has been proposed for management communication between the DSLAM and ATU-R. The Common Object Request Broker Architecture (CORBA), used to interconnect various management systems and entities, is also relevant here.

3.9.2.1 The Simple Network Management Protocol (SNMP)

Today, SNMP is the most widely deployed protocol. SNMP is well understood, relying on Management Information Bases (MIBs) which describe the various parameters which may be configured or monitored within a system. This is a virtual information store, defined via the mechanisms described in the Structure of Management Information (SMI). A given device will include a set of standard and vendor-specific MIBs, the first set resulting from work in organizations such as the IETF, ATM Forum, and ADSL Forum, while vendor MIBs relate to device-specific parameters and capabilities. SNMP operates via a limited set of commands between the management platform and the managed device. These include

- get—used to retrieve specific management information.
- get-next—used to retrieve, via traversal of the MIB, management information.
- set—used to alter management information.
- trap—used to report extraordinary events.

For example, an operator requiring device status information will enter a command via a management GUI. This will then trigger an SNMP Get to the appropriate MIB. In the same way, if a device parameter is to be changed, the station will generate an SNMP Set. Finally, the network element will maintain a list of counters and thresholds, set by the user. If one of these thresholds is exceeded, an SNMP Trap will be sent to the management station.

SNMP consists of an overall architecture described in RFC-2271, as well as mechanisms for describing and naming objects and events, protocols for transferring management information, protocol operations for accessing management information, and a set of fundamental management applications. The method of object naming is known as the Structure of Management Information (SMI), first described in RFC-1155 as SMIv1 and then later updated in RFCs 1902, 1903, and 1904 as SMIv2. The actual message protocol was first described in RFC-1157 as SNMPv1. SNMPv2c, described in RFCs 1901 and 1906 was never widely deployed and is not a standards-track protocol. RFC-1157 and 1906 also described the actual protocol for accessing the management information as well as the proper PDU formats. This was due to problems in reaching closure on security issues. The more recent SNMPv3 as defined in RFCs 2272 and 2274 will see wide use in the future. Finally, the applications as described in RFC-2273 are designed to initiate, respond to, or forward SNMP requests.

Another management protocol with the same basic intent as SNMP is the Common Information Management Protocol (CMIP). Although many service providers have supported this protocol in the past, the direction of the industry is increasingly toward SNMP, even in the public management space.

3.9.2.2 Web-Based Architectures

Although SNMP is still the most widely deployed management protocol, the use of web-based architectures is becoming increasingly important and will play an even greater role in the future. Here, the network element is equipped with an HTTP server, while a user on any common browser has the ability to configure or monitor the device. No longer is a separate management application required. On its own, this technique is useful for on-site configuration or troubleshooting. When integrated with a database and functionality such as that available with Java, a web-based management environment may begin to replace much of the capabilities currently available via SNMP platforms.

3.9.2.3 The Simple Device Protocol (SDP)

SNMP, CMIP, and web-based management tools all share one thing in common—they require IP connectivity among the managed devices. If the managed device has no inherent need to support IP-layer intelligence, an ATM-layer management interface would dispense with the need for this additional complexity. This could be especially important if a DSLAM is constructed with distributed line card shelves. Thus the premise behind the Simple Device Protocol (SDP), which uses ATM for all device control and configuration. SDP establishes well known VPIs/VCIs for management communication, with messages carried directly over the ATM Layer. The level of functionality is expected to be equivalent to that of G.997.1, described in section 3.9.3.

3.9.2.4 The Common Object Request Broker Architecture (CORBA)

Irrespective of the management protocol in use, be it SNMP, CMIP, or Web, the ultimate requirement is that open interfaces are provided to existing, businesswide management systems. An early example of this is the "Q3" interface described in many service provider management documents. More recently, the Common Object Request Broker Architecture (CORBA) (http://www.omg.org/corba/) has gained favor as a way to interconnect management systems from multiple vendors. A CORBA environment may be visualized as a bus (Figure 3.26) connecting the various element and network management platforms with multiple applications. These applications, written in different programming languages and residing on different operating systems, communicate via a set of APIs. The CORBA also includes a set of utilities to assist in building interoperable applications.

FIGURE 3.26.

CORBA

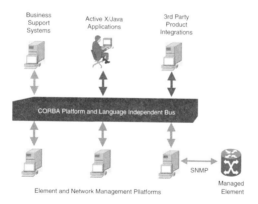

3.9.3 Element Layer Management—G.997.1 and the ADSL Line MIB

Relating the concepts introduced in the previous section to actual implementations, G.997.1, under development within both the ITU-T and the ADSL Forum, defines the various parameters that may be managed between the ATU-C and ATU-R. It accomplishes this by defining an Embedded Operations Channel (EOC) at the physical layer, connecting

ADSL Management Entities (AMEs) located in the two devices (Figure 3.27) and possibly connecting into an external management system via the "Q" interface. Capabilities of G.997.1 include configuration, fault, and performance management. The EOC, known as a Clear EOC, carries the SNMP traffic in a full-duplex HDLC-framed channel whenever the link is in normal operating mode (4 kbps minimum: no guarantee, retransmission, or acknowledgment).In the context of the OAM flows, the ADSL link at the physical layer relates to F1–F3, although the recommendation defines functionality associated only with the F3 (transmission path) flow. Looking more specifically at ADSL, the standard introduces the concept of an ADSL Line and ADSL ATM Path, the former connecting the digital outputs of the respective ADSL modems, while the latter extends from the ATM interfaces at the ATU-C and ATU-R.

FIGURE 3.27.

G.997.1

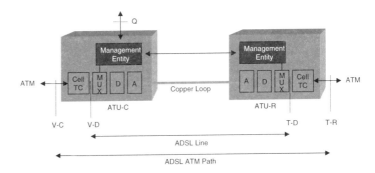

The ADSL Line MIB takes the concepts defined within G.997.1 and outlines an actual SNMP MIB. The MIB contains objects describing the physical state of the line such as upstream and downstream bandwidth, the number and types of faults over time, and any preset variables such as noise margins. The appendix outlines the major components of this MIB, which reflects the types of statistics that are relevant across the ADSL loop.

Note that the existing ADSL Line MIB is modem-independent, necessitating additional MIBs for CAP and DMT-specific parameters. This state of affairs was not the original intent of the ITU-T, however, and is subject to change. In addition, a CMIP implementation of G.997.1 is under definition as well, though this is expected to be less widely implemented.

3.9.4 Network Layer Management

Moving a layer up from element management (see Figure 3.25), operators have a requirement to monitor and troubleshoot network-layer parameters. This monitoring may take place between two devices, such as ADSL modems and the link connecting them, or it may extend across an ADSL network. An adequate view of the network will require input from multiple network elements in the path, feeding into a central management station. In addition, this view should be available at different layers of the protocol stack. For example, the ADSL transmission provider may be interested only in ADSL and ATM layer connectivity between the user CPE and the DSLAM or may wish to extend this monitoring across an ATM network to the PVC termination point. Conversely, the actual ADSL end-user service may be provided by an ISP. Here, the ISP requires IP Layer and even session layer (that is, higher-layer protocols and applications) monitoring. One area in which work is still to be done is the handoff of management information between different organizations. For example, if the ADSL transmission provider realizes that an ATM link has failed, the ISP should be notified as an aid in troubleshooting. This is still a difficult task.

3.9.5 Subscriber Provisioning

Separate from the requirement to configure and monitor the various network elements from a device-centric view is the need for a more service-centric view based on subscribers. For example, a network operator provisioning a new user with a given set of service characteristics should be able to enter a set of parameters that will then trigger the necessary element configurations across the network. In its simplest form, *user provisioning* is deployed on a box-by-box basis. More advanced systems support what is known as *flow-through-provisioning*, either across the DSL components of a network or even including an ATM core network. These are functions that occur on the Service Management Layer of the TMN architecture.

Figure 3.28 depicts an example of a typical provisioning system, whereby a service request is first associated with a service description (templated ahead of time).

FIGURE 3.28.

Flow-through Provisioning

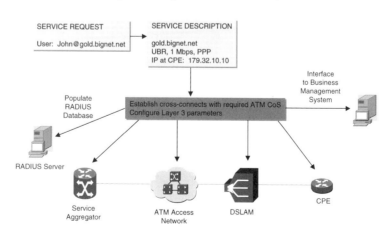

This then feeds into an agent that generates the actual device configurations based on a knowledge of available resources. This may include use of SNMP, CMIP, and/or web-based techniques depending on the device in question. The system may also interact with the provider's AAA system (i.e., RADIUS), automatically populating the server with the subscriber's authentication and authrorization information. Finally, interfaces will be provided to the provider's various business management applications, including billing and SLA tracking. Note that the actual protocol used to connect the various management entities across the network may be CORBA, described in section 3.9.2.4. The one major complexity with flow-through provisioning is whether a single entity controls all the devices along the end-to-end service path. In fact, this is not usually the case, with the ILEC or IAP responsible for the access network (DSLAMs) and the ISP responsible for the aggregation function. This introduces (solvable) challenges in enabling true end-to-end provisioning.

Now looking at user provisioning, consider a system offering a single view into the network for the configuration of DSL subscribers. This system will have interfaces into the DSLAMs, aggregators, core switches, and even CPE, if required (realizing that the CPE could be configured via the DSLAM). To understand how such a provisioning architecture would work, the concept of data models and user templates is introduced. Data models relate to the types of services offered across the DSL network, such as end-to-end PVCs, bridging, and PPP/L2TP tunneling. These are described in greater detail in Chapter 4.

Within each service, the network administrator may predefine a set of templates corresponding to the service parameters. For example, a PPP/L2TP "gold" service may be defined at a set bandwidth per subscriber and Layer 3 CoS. Figure 3.29 details this scenario, with a new subscriber associated with a given service class among a number of predefined templates. This information is then used to generate a subscriber record used in automatic device configuration, containing parameters such as the ATM CoS, the type of service, the proper VPI/VCI combination to use, and any additional service-enabling information such as the address of the RADIUS server.

The network administrator, upon receipt of a service request, will now configure the necessary Layer 2 and Layer 3 parameters across the network. In the case of PPP/L2TP, this Layer 2 configuration will include setting up the PVC from the subscriber CPE, across the DSLAM, and into the aggregator where the PVC terminates. At both the CPE and the aggregator, the user's PPP session must be configured, including IP addressing (if no DHCP) and security mechanisms. In the case of L2TP tunneling, the administrator also defines the tunneling parameters between the aggregator and the Home Gateway terminating the tunnel. In the example shown in Figure 3.29, a "gold" tunnel is defined. Any subscribers with this service profile will share this tunnel.

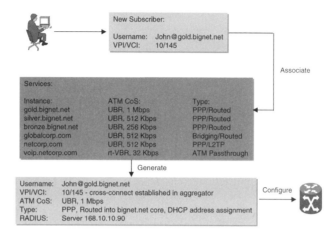

FIGURE 3.29.

Subscriber Provisioning

Revisiting the earlier discussion of MIBs, the use of SNMP is not limited to the element layer. For example, one vendor has implemented a VPN MIB providing information on tunneling between a NAS/LAC and a HGW/LNS. Managed entities include the global VPN state, per-tunnel state and user information, and the failure history per user. When troubleshooting a VPN implementation, this type of information may prove critical.

3.9.6 Authorization and Directory Services

Operating on a different plane from the element, network, and subscriber provisioning systems already described are the mechanisms which allow already provisioned users to gain access to network resources. Actions here include populating a database with the necessary accounting, security, directory, and QoS attributes to identify uniquely individual subscribers and their permissions. The requirements here are no different than those found within Internet dialup services: When users first log into the network, their identity is validated against an authentication and authorization database. Upon validation, billing commences (if based on time), and the server may also return what services the user has and does not have access to.

In the following sections, RADIUS and TACACS+ are first detailed, leading into a description of Directory Enabled Networks (DEN)and Global Roaming.

These various protocols and servers all fill a need within an ADSL service offering. The RADIUS or TACACS+ server is where ADSL providers, ISPs, and corporations maintain user identity and accounting information, critical for security and billing. The UCP integrates the RADIUS server with additional provisioning components, while global roaming manages remote connectivity. In the near future, the DEN initiative will provide a central-ized repository for network services.

3.9.6.1 RADIUS

An essential component of any Internet access architecture is user authentication and accounting. This requires a server capable of passing user access parameters to the access server (dialup or DSL) as well as tracking connection time. The *Remote Access Dial-In User Service (RADIUS)* server, developed as part of the Internet dial architecture, performs these functions. This protocol was originally developed by Livingston Enterprises but is now in the public domain, documented in two RFCs: RFC 2058, which is the specification, and RFC 2059, which details the optional accounting standard. The RADIUS server relies on a client-server model, with the PPP termination point as a RADIUS client, which passes user information to a designated server, and then acts on the response. The RADIUS server provides services to one or more clients and is usually a dedicated system. One powerful feature of RADIUS is the capability of vendors to add extensions to the basic protocol to meet the needs of their hardware and/or their ISP partners. As an example, the parameters that allow for tunneling (L2F, L2TP) are optional.

Communication between the client and server is via UDP as depicted in Figure 3.30.

Figure 3.30.

RADIUS Exchanges

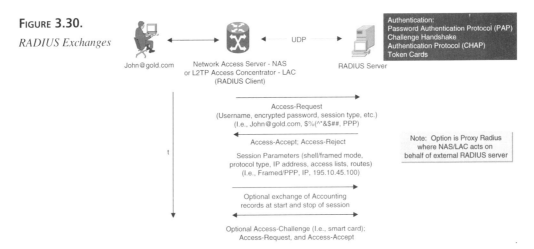

Under operation, when a user logs on, the client passes the username, password, and LAC-specific information to the server via an Access-Request message. Authentication methods may include the Password Authentication Protocol (PAP) or token cards. The server then proceeds to search its database for the username. If successful, the server returns an Access-Accept along with a list of attribute-pairs describing the parameters of the session. These parameters may include the protocol type, IP address for the user, access lists, and routing. Now looking at accounting, typical parameters include time, packets sent/received, and bytes sent/received. Note that in the context of ADSL, under existing tariffing models, accounting is less relevant for a local ISP, although it may become very important in a VPN architecture where users access their home ISPs via third parties. Chapter 6 depicts the use of RADIUS at the LAC as part of the configuration examples.

3.9.6.2 TACACS+

The Terminal Access Controller Access Control System Plus (TACACS+) is the latest version of a user authentication protocol that has existed since the early 1980s in the form of TACACS and XTACACS. Like RADIUS, TACACS+ includes authentication, authorization, and accounting, tying each service to a potentially independent database. This enables the implementation of different mechanisms for each of these functions. For example, KERBEROS may be used for the authentication phase, and authorizations may be requested once a session has been established. TACACS+ also relies on TCP for guaranteed communication between the client and server, also useful in determining whether a server has gone down. Note that RADIUS emulates this via variables controlling retransmit attempts and timeouts. Next, TACACS+ encrypts all transactions via MD5, unlike RADIUS which encrypts only the password in the Access-Request packet. Thus accounting and services attributes are not subject to eavesdropping. The TACACS+ protocol also natively supports multiple Layer 3 protocols such as the NetBIOS Frame Protocol Control and AppleTalk Remote Access (ARA). Finally, accounting may contain locations, protocol, and even commands invoked in addition to the usual start-stop records. Although TACACS+ is the more powerful AAA mechanism, the majority of ISPs have deployed RADIUS.

3.9.6.3 Directory Enabled Networks and the Active Directory

As directory systems evolve, the UCP, along with other user provisioning systems, will evolve to make use of them. The most interesting initiative is the Directory Enabled Network (DEN). This architecture merges X.500 concepts, the Common Information Model from the Distributed Management Task Force, and internetworking architectures. It creates a central repository of all available network elements and services, including individuals, devices, policies, protocols, locations, and applications. Applications in areas as diverse as billing, provisioning, service creation, authentication, and IP address management will draw on this information to provide a common view of network resources. The net result is that the use of network resources, such as bandwidth, may be optimized by enabling policy-based management.

Probably the best way to understand the use of DEN is through a practical example. Consider a human resources (HR) manager logging into the network to update an employee's records. First, the AAA server contacts the Active Directory to authenticate the HR manager. Since the manager is accessing an HR application, a policy in the directory indicates that IP security (IPsec) is required for the session. The AAA server now establishes the encryption. Once online to the HR application, the manager updates the records, resulting in a trigger to give the promoted employee preferred network access and server privileges.

One example of a DEN implementation is Microsoft's Active Directory (AD), available within both the NT and Unix environments. AD as well as other DEN implementations rely on the Local Directory Access Protocol (LDAP) as their data access protocol. To enable easier interfacing to the actual directory, the Active Directory Service Interfaces (ADSI) has been created. This is a set of open interfaces that allow developers to access LDAP as well as other directory systems.

3.9.6.4 Global Roaming

Because the majority of ISPs will not deploy their own ADSL network (as they do not control the copper loop), an infrastructure that will allow any ADSL subscriber access to his or her home ISP is required. Note that this requirement also exists in the dial space because it is not cost-effective for the average ISP to deploy PoPs globally. A Global Roaming Server allows customers of one ISP access via the facilities of another. For example, a subscriber's residence may be in Miami, and the ISP of choice is the local `miami.net`. Since `miami.net` focuses on South Florida, it cannot be expected to deploy PoPs throughout the United States. However, when the subscriber travels to San Jose, for example, he requires access to `miami.net`. Currently, this is accomplished via 800 numbers, although this scheme scales only so far. Using the GRS, the subscriber may dial in to the facilities of `sanjose.net`, the local ISP. The GRS then proxies for the subscriber's identity at `miami.net`'s AAA server and, if valid, establishes a secure tunnel to `miami.net` (see Figure 3.31).

The relationship between `miami.net` and `sanjose.net` is handled by the GRS at `sanjose.net` and the AAA server at `miami.net`. This service is even more critical in the ADSL space, where the equivalent of 800 connectivity cannot exist. The concept of the GRS forms the basis for one of the more critical ADSL service models, described in Chapter 4.

FIGURE 3.31.

Global Roaming

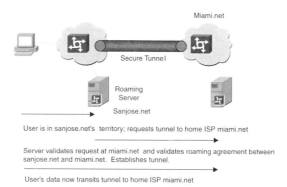

User is in sanjose.net's territory; requests tunnel to home ISP miami.net

Server validates request at miami.net and validates roaming agreement between sanjose.net and miami.net. Establishes tunnel.

User's data now transits tunnel to home ISP miami.net

3.10 Regulatory Considerations

Although the components described in the previous sections comprise the physical infrastructure of the ADSL system, where and how service providers may actually deploy them is influenced in a great way by the local regulatory environment. The word "local" is critical here, since laws vary country by country and sometimes even on a regional level within a country. In the United States, the Telecommunications Deregulation Act of 1996 sets the stage for the ways in which the traditional phone companies (known as Incumbent Local Exchange Carriers, or ILECs) and the newer alternate carriers and ISPs (known as Competitive Local Exchange Carriers, or CLECs) may deploy services. The intent of the law is to open up the ILECs' facilities with a goal of encouraging competition and making advanced services available at lower costs. The ILECs must price these unbundled elements (that is, copper pairs and basic ATM service) on a wholesale basis to both the CLECs and their internal value-added arms. If the ILECs fail to comply, the state utility commissions are expected to intervene. In Europe, PTT deregulation and privatization exert the same influence. The environment is fluid, and what may be illegal one year may be perfectly legal the next.

3.10.1 United States and Telecommunications Deregulation

In the United States, the 1996 Telecommunications Deregulation Act defines which services the ILECs are permitted to offer and how they must provide equal access to their infrastructure for the CLECs. For example, a Regional Bell Operating Company (RBOC) will normally have two totally different business entities: an ILEC component that controls the voice switching and transmission and provides basic phone service, and a CLEC offering Internet services. The ILEC is not permitted to offer these higher layer services, while the

CLEC must purchase bandwidth and even copper access from the ILEC. Other CLECs, not associated with the RBOC, are in theory given the same level of access.

Within this framework, CLECs have a number of options for hardware and service deployment. The simplest is to procure space in the vicinity of the ILECs CO, backhauling all traffic, as shown in the first option in Figure 3.32. Although this is simple in concept, it does have a major impact on ADSL in particular, since the technology is so dependent on loop length and interactions within cable groups.

FIGURE 3.32.

Co-location
Alternatives

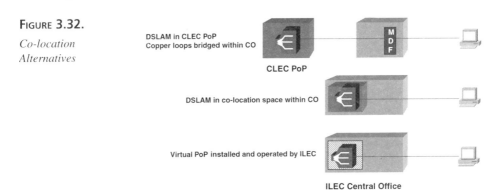

When considering alternatives, most think of physical colocation as the solution (option 2 in Figure 3.32). Here, the CLEC leases space within the ILEC's premises. This provides best access to the loop, with tradeoffs being a high up-front cost and the necessity to comply with the CO requirements in terms of physical profiles, power consumption, and available space. Yet another alternative is known as Virtual Colocation, where the CLEC contracts to the ILEC for service installation and maintenance (option 3 in Figure 3.32). Where the CLEC cannot justify dedicated hardware, this may be a viable solution. However, it is potentially the most expensive. An interesting point is that although the deregulation act requires the ILECs to unbundle the copper loops, they have no requirement to do anything in the DLC space. This in fact is a way for them to retake control of the local loop.

Now mapping this environment to ADSL hardware, the ILEC, controlling the copper loops, will normally own and operate the DSLAM. They may also provide the user's CPE, although in most cases the carrier's CLEC component will supply the ADSL modem or ADSL connected router to the customer. A portion of the cost of this device is usually charged to the user at the time of ADSL installation. The degree to which they manage this device will depend on the hardware deployed. The ILEC will also normally control the ATM access network if one exists, and if ATM, the core as well. In contrast, the CLEC will control a routed core, user provisioning, any servers or caches, and if offering a managed router service as part of ADSL, the CPE as well.

One reason for separating these functions is that personnel in both the ILEC and the CLEC may have different types of training: The first group is better versed in transmission and switching, while the second has routing expertise. The aggregator is unique, in that it may be operated by either the ILEC or the CLEC depending on the services deployed. For example, if used for generic ATM switching or possibly PPP tunneling, it may fall under the domain of the ILEC; if routing is activated, the CLEC will have control. A non-RBOC CLEC may deploy a DSLAM if it has access to the copper pairs and colocation rights. These are two very important points, since the ILEC may not tariff dry copper, as it is known in the industry, and if it does, regulations governing the mix of traffic types on the copper bundle may be too restrictive. The ILEC need only say that ADSL will have a negative impact on existing T1, HDSL, or even ISDN services, as detailed in Chapter 2. Colocation may also be a sticky point. If space is not available, or if colocation terms cannot be met, the CLEC may be forced to locate the DSLAM in a remote site. This presents problems with ADSL since the cable length is being extended, consequently reducing the available bandwidth. In addition, between the CLEC's location and the ILEC CO, the greater bandwidth is now pointing toward the CO, precluding use of the cable bundle for ADSL services outward from the CO.

3.10.2 European Regulatory Environment

The regulatory environment is a bit different in Europe and varies on a country-by-country basis depending on the state of privatization. In those countries where the government has divested the PTT, the degree to which third parties may offer ADSL services will depend on whether dry copper is available and whether there is a history of competition for basic services. Even where the incumbent telco is in a position to offer ADSL services, it will still backhaul to one of a number of independent ISPs. The one major architectural difference between U.S. ILEC and European telco deployments is at the CPE. Whereas the ILECs in the United States commonly handoff CPE responsibility to another group, European telcos have this concept of an active NT, where they provide the line termination. This is analogous to the difference in ISDN installations between the two sides of the Atlantic. In the United States, the provider simply terminates the copper at a wall jack, European installations include a powered NT for manageability. Extending this architecture to ADSL, we expect to see a number of installations in Europe where the incumbent telco provides the basic ADSL modem, with the user then providing a separate router when required. Moving to those countries where the government still controls the PTT, the major difference is that there is no possibility for third parties to enter the ADSL transmission business. In fact, in more than a handful of nations, the government even controls the single ISP, usually an outgrowth of the PTT. Interestingly, even in those countries where there is no competition to the national PTT, there is still demand to deploy ADSL.

3.10.3 Pacific Rim and Other Regulatory Environments

Within Hong Kong, Japan, Taiwan, Singapore, and even China, deregulation is underway, with CLECs coming into existence to serve business subscribers and international connectivity. As with North America and Europe, many of these draw on alternative infrastructures such as power and rail. However, the unbundling of the local loop in this region is still in its infancy, with many countries not yet setting a date for this to occur. Thus the CLECs will operate at the boundaries of the traditional telco, terminating any ADSL VCCs presented. Where permitted, the CLECs may also enter into colocation arrangements as described in section 3.10.1, "United States and Telecommunications Deregulation."

In other parts of the world, deregulation is occurring in the same way on a country-by-country basis. First, the government opens the market to competitive business service providers and ISPs, then opens the voice market to competition. The last stage of deregulation is the unbundling of the loop. In the United States, this last stage is really a two-edged sword for the new providers, with the incumbents usually countering by advancing their DLC plans. The DLC environment is traditionally not subject to unbundling.

Endnotes

ADSL Forum, ADSL Network Element Management, TR-005, March 1998.

ADSL Forum, "Interfaces and System Configurations for ADSL: Customer Premises," Technical Report TR-007, March 1998.

ADSL Forum, SNMP-based ADSL Line MIB, TR-006, March 1998.

ADSL Forum, ADSL Network Management Framework, WT-024, June 1998.

ADSL Forum, CAP Line Code Specific MIB, Network Management Working Text WT-023, September 1998.

ADSL Forum, DMT Line Code Specific MIB, Network Management Working Text WT-022, September 1998.

ADSL Forum, CMIP Specification for ADSL Network Element Management, Working Text WT-025, October 1998.

Arang, M., Dugan, A., Elliott, I., Huitema, C., and Pickett, S., "Media Gateway Control Protocol, Draft 0.1." draft-huitema-MGCP-v0r1-01.txt, November 1998.

ATM Forum Technical Committee, "Physical Interface Specification for 25.6 Mb/s over Twisted Pair Cableaf-phy-0040.000," November 1995.

Bathrick, Definitions of Managed Objects for the ADSL Lines, draft-ietf-adslmib-adsllinemib-01.txt, August 1998.

Blumenthal, U. and Wijnen, B., (RFC-2274) User-based Security Model (USM) for version 3 of the Simple Network Management Protocol (SNMPv3), January 1998.

Carrel, D. and Grant, L., "The TACACS+ Protocol Version 1.78," Internet Draft, draft-grant-tacacs-02.txt, January 1997.

Case, J.D., Fedor, M., Schoffstall, M.L., Davin, C., (RFC-1157) Simple Network Management Protocol (SNMP), May 1990.

Case, J., McCloghrie, K., Rose, M., and Waldbusser, S., (RFC-1901) Introduction to Community-based SNMPv2, SNMPv2 Working Group, January 1996.

Case, J., McCloghrie, K., Rose, M., and Waldbusser, S., (RFC-1902) Structure of Management Information for Version 2 of the Simple Network Management Protocol (SNMPv2), January 1996.

Case, J., McCloghrie, K., Rose, M., and Waldbusser, S., (RFC-1903) Textual Conventions for Version 2 of the Simple Network Management Protocol (SNMPv2), January 1996.

Case, J., McCloghrie, K., Rose, M., and Waldbusser, S., (RFC-1904) Conformance Statements for Version 2 of the Simple Network Management Protocol (SNMPv2), January 1996.

Case, J., McCloghrie, K., Rose, M., and Waldbusser, S., (RFC-1905) Protocol Operations for Version 2 of the Simple Network Management Protocol (SNMPv2), January 1996.

Case, J., McCloghrie, K., Rose, M., and Waldbusser, S., (RFC-1906) Transport Mappings for Version 2 of the Simple Network Management Protocol (SNMPv2), January 1996.

Cisco Systems,"Cisco IOS Technologies: RADIUS Support in Cisco IOS Software", http://www.cisco.com/cpropart/salestools/cc/cisco/mkt/ios/tech/security/rdius_wp.htm, 1997.

Data Communications, April, 21, 1998, and ADSL Forum Tutorial http://www.adsl.com.

Drout, D., Underwood, J., and Adams, P., "NORM—Provide Customer Service," ADSL Forum 97-117, August 1997.

3

Handley, H., Schooler, E., and Schulzrinne, H., "Session Initiation Protocol (SIP)."

Harrington, D., Presuhn, R., and Wijnen, B., (RFC-2271) An Architecture for Describing SNMP Management Frameworks, January 1998.

Harrington, D., Presuhn, R., and Wijnen, B., (RFC-2272) Message Processing and Dispatching for the Simple Network Management Protocol (SNMP), January 1998.

Home Phoneline Networking Alliance, Universal Serial Bus Specification, Revision 1.0, January 1996, www.usb.org., A White Paper, June 1998, www.homepna.com.

IEEE Std 1394-1995, Standard for a High Performance Serial Bus, 1995.

"IPv4 over IEEE 1394," Internet Draft, draft-ietf-ip1394-ipv4-09.txt, June 1998.

ITU-T Recommendation H.323, "Visual Telephone Systems and Equipment for Local Area Networks Which Provide a Nonguaranteed Quality of Service," February 1998.

ITU-T, Physical Layer Management for ADSL Digital Transmission Systems, Draft ITU-T Recommendation G.ploam (AB-009R1), SG-15, 1999.

Levi, D., Meyer, P., and Stewart, B., (RFC-2273) SNMPv3 Applications, January 1998.

Orth, B., ADSL: A Telecom Operator Perspective, presentation to the ADSL Symposium, CeBit, 1998.

Paton, C. and Adams, P., "NORM—Repair and Maintain Service," ADSL Forum 97-116, August 1997.

Rose, M. and McCloghrie, K., (RFC-1155) Structure and Identification of Management Information for TCP/IP-based Internets, May 1990.

Speakman, T., Farinacci, D., Lin, S., and Tweedly, A., "PGM Reliable Transport Protocol Specification," Internet Draft: draft-speakman-pgm-spec-02.txt, August 1998.

Texas Instruments, "Data Transmission, Part 3, High Speed Standards," March 1998.

Wahl, M., Howes, T., Kille, S., RFC 2251 - Lightweight Directory Access Protocol (v3), December 1997.

Wilhelm, The Simple Device Protocol (SDP), ADSL Forum Contribution 98-158, September 1998.

Chapter 4

Services

With an understanding of the basic ADSL infrastructure, protocols, and devices, the next step is combining these diverse elements into end-to-end service models enabling higher layer applications. Figure 4.1 depicts this taxonomy by outlining models based on end-to-end ATM VCCs, bridging, routing, and PPP. The PPP alternatives further segment into those that support VPNs (such as PTA, L2TP, multidestination via NAT, and MPLS) and a service model that does not support VPNs, based on termination into routing.

It is important to note that an ADSL provider may offer one or a number of these services, and that some services may be enabled by the provider of the basic ADSL service while others are enabled at the upstream ISP or corporate gateway. Therefore, there is in fact some overlap in the service models.

For example, the ADSL transport provider may provide PVCs, while the ISP connecting to the transport service terminates these PVCs and provides end-to-end bridging or routing services to its subscribers. Related to all of these models is a method by which a user may easily access different destinations, services, and content. The remainder of this chapter looks at these end-to-end models in detail, focusing on the hardware devices, protocols, and data flows. Chapter 5,

FIGURE **4.1.**

Service Taxonomy

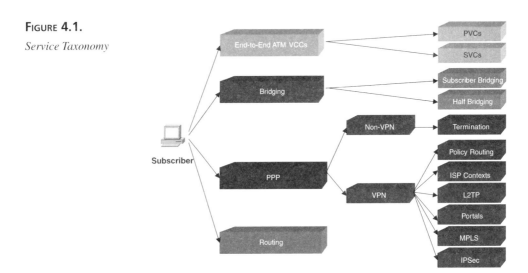

"Applications," then takes a subset of these models and provides complete configuration scenarios—basically a cookbook to service deployment.

4.1 End-to-End ATM Virtual Circuit Connections

Looking back at the reference architecture presented in Chapter 2, recall that the ADSL local loop relies on ATM encapsulation between the CPE (ATU-R) and the DSLAM (ATU-C) in the Central Office. Also remember that the DSLAM presents an ATM trunk interface into the ATM access or regional network. Thus the simplest of ADSL service models, although not lacking in advanced capabilities, relies on establishing ATM PVCs or SVCs from a subscriber to a destination.

This topology is in fact the most common within early production ADSL deployments, since the transmission group providing the ADSL service is usually well versed in ATM switching, while the upstream entities such as the ISPs are versed in routing and higher layer services.

These ATM virtual circuits will usually carry internetworking traffic, terminating at Layer 3 entities at either end of the circuit. In the scenario shown in Figure 4.2, the PC or CPE router encapsulates the data traffic via RFC-1483 bridging or routing, RFC-1577 for IP if the network supports SVCs, LANE or MPOA if in the campus (or possibly MAN), or PPP in most cases over ADSL. It then passes the encapsulated data into the ATM Layer which adapts IAW AAL5 and segments the data into ATM cells.

Depending on the QoS requirements of the original data, the ATM PVC or SVC may conform to VBR-rt, VBR-nrt, ABR, GFR, or UBR. Alternatively, the CPE may support native ATM services such as voice or video. In the former case, the CPE implements the ATM CES or VTOA, adapting the traffic IAW AAL1 or AAL2 before generating the cells and passing the traffic onto a CBR or VBR-rt PVC or SVC.

Video across ATM usually relies on MPEG-2 and either AAL1 or AAL5 (preferred). Voice and video over ATM were briefly introduced in Chapter 2, "Architecture," along with the different methods of transporting data across ATMs. Figure 4.2 depicts these encapsulations for both PVC and SVC-capable CPE and aggregators.

FIGURE 4.2.

ATM PVC and SVC Protocol Encapsulations

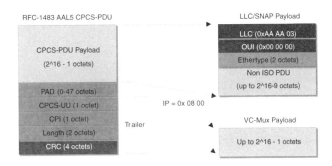

This ATM-centric architecture has both advantages and disadvantages in terms of service possibilities, provisioning, and resource use.

4.1.1 Advantages and Disadvantages to ATM

To its advantage, this ATM-centric model makes the ADSL local loop look like just another media for multiservice ATM delivery. For example, a user may establish multiple VCCs, each with a different QoS. Data then traverses one VCC, while voice (via VTOA) uses another. This may be a reasonable architecture for a large organization, where the headquarters and regional sites connect via "native" ATM at OC-3/STM-1 or DS3/E3, while the branches use ADSL for access.

A second advantage of ATM is its support of Layer 2 VPNs. Given proper network management, the mesh of VCCs serving an organization guarantee security. For those uncomfortable with the concept of high-layer security such as that based on Layer 3 VPNs via L2TP, MPLS, or IPSec, this is an alternative.

However, relying on end-to-end VCCs does present some difficulties in terms of managing PVCs, resource allocation, and if implementing SVCs, security and accounting. Note that these various issues, described in the sections that follow, are no different than those encountered within any large ATM deployment.

4.1.2 PVCs

In an end-to-end PVC environment, the ADSL provider provisions VCCs between each sub-scriber and a corresponding upstream destination. Since most ADSL CPEs support multiple VCCs over the loop, nothing prevents the provisioning of more than one PVC at the sub-scriber site. A typical configuration may include corporate telecommuters connected over the ADSL transport network to their corporate gateway or branch offices connected to an ATM-connected headquarters site. In both instances, the ADSL provider plays no role in higher-layer services except to ensure that the ATM Layer, including the CPE, DSLAMs, and ATM switches, adequately supports any applications requiring QoS.

Figure 4.3 depicts the scenario where the CPE segments user data into ATM cells, transports these cells across the ADSL network, and then reassembles them at their ultimate destination.

FIGURE **4.3.**

PVC Connectivity
Events

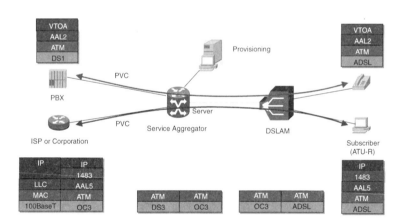

The problem with relying on PVCs relates to provisioning. Unlike existing PVC-based ATM deployments, with maybe hundreds of termination points, ADSL deployments must support tens of thousands of subscribers. And, unlike many current ATM deployments, the churn is expected to be greater.

Consider a subscriber first connecting to one ISP and then to another. The ADSL provider must reprovision the PVC when this occurs, incurring additional costs. This is one element leading to a demand for the Internet wholesaling model based on PPP tunneling, described in the next section. Therefore, even with scalable provisioning tools, management becomes an issue.

An alternative viewpoint is that Frame Relay networks have already successfully addressed these scalability issues. The next issue is with the connectivity itself. PVCs allow only point-to-point connections, so in most cases, all traffic flowing from one point on the network to another, even if multicast, must pass through a hub site. This may or may not be an ideal situation for telecommuter or branch connectivity, with the tradeoff between the number of PVCs required and inefficiency in data flows.

Under the PVC architecture, the ADSL provider establishes one or more VCCs from the subscriber. Initially, these will be UBR due to traffic management limitations on the DSLAMs and CPE, but as mentioned previously, nothing precludes the provisioning of CBR, VBR, GFR, and even ABR in the future. In this case, both ATU-R and DSLAM must support buffering and queuing capable of differentiating between the different ATM service categories.

The VCC is now shaped to the user's ADSL traffic contract, less than or equal to that of the physical capabilities of the loop after rate adaptation. It passes through the DSLAM, which multiplexes the VCC onto an OC3/STM1, DS3, or DS1 IMA ATM uplink with traffic from other subscribers. Depending on the service offering, the provider may oversubscribe this trunk to a greater or lesser extent.

If the service is UBR, this does not present a problem since UBR assumes a data connection with higher-layer congestion avoidance (such as that provided by TCP). If the service is one of the other traffic classes, the DSLAM ensures priority for the CBR and VBR traffic. Once again, this is a provisioning issue.

In most cases, the user VCCs enter the DSLAM with an arbitrary VCI but with the VPI set to "0." This does not present a problem, because the DSLAM assigns each user a unique port on the device. The DSLAM now remaps the VPI based on the user's port number. For example, a user on port 4 will have its VCCs set to VCI=x, VPI=4.

Note that this type of mapping allows only a total of 255 users across the DSLAM uplink, an unnecessary limitation. An alternative is to use the complete VCI field for mapping both the user's VCCs and the user's unique identification number (set via management) across the trunk since no ADSL user is expected to require more than about 32 VCCs. Another approach is just to use the VPI and VCI field as a contiguous space, with no direct relation between either field and the user number.

The VCCs now enter an ATM access or regional network, where they pass through one or more switches before reaching their destination. At each point, the switches remap the VCI and VPI fields in the cell headers, although this is transparent to the end user.

If the connection contains a "hard" PVC, the connection follows a predefined path across the network, susceptible to failure if a link is taken out of action. Preferable are soft-PVC connections, predefined only between the CPE and the first hop ATM switch and following any available path across the network as determined by the PNNI.

SVCs, of course, rely on the PNNI for routing at the time of circuit establishment. An optimization of this architecture is to group VCCs destined for a single upstream destination into a VP for transport across the backbone. This eases provisioning.

4.1.3 SVCs Switched Virtual Circuits

SVCs are a step up in complexity and functionality and rely on signaling support (as described in Chapter 2) on the CPE, on the DSLAMs, and within the ATM transport network. Note that the soft-PVC approach, described in the previous section, also exists. Within an SVC deployment, the subscriber signals on-demand a VCC to an upstream destination or service. Depending on the application, this SVC may be CBR, VBR, ABR, GFR, or UBR.

Note that the QoS requirements of both native ATM and internetworking applications may trigger an SVC with a specific ATM service category. In this case, the ATM interface on the CPE and/or the ATU-R must support ATM TM as well (as was the case with PVCs).

Take a look at the flow of events shown in Figure 4.4. The CPE initiates a Sig 4.0 request across the ADSL loop (the UNI) to the DSLAM. The DSLAM interprets this request and forwards it across its PNNI trunks to an aggregator or ATM switch. The last switch in the path then forwards the request to the destination, which accepts or rejects the call. Even though this discussion assumes that the DSLAM is SVC-capable, there are two alternative approaches based on the Virtual UNI or on the VB5.2 architecture, both described in Chapter 2.

FIGURE **4.4.**

SVC Connectivity
Events

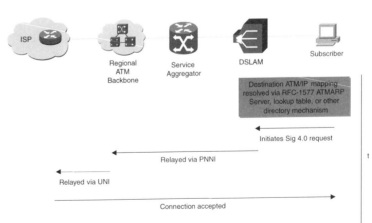

Concerns in deploying SVCs include, naturally, the requirement that all devices along the path support SVCs. This is not a given in the first phase of ADSL deployments but should become more commonplace during the latter half of 1999. In addition, the provider must support the necessary accounting applications for usage-based tariffing. If the SVCs are within an internetworking environment, where accounting via Layer 3 AAA mechanisms are a possibility, this may not be a problem. However, one can expect that in native applications cell- and duration-based accounting will be a requirement.

Finally, the provider must ensure that ATM layer security is in place. Options here for segmenting user connectivity include Closed User Groups (CUGs) and NSAP address access lists, in anticipation of the more complete ATM security architecture.

4.2 PPP Point-to-Point Protocol

Although ATM VCCs are a starting point in terms of service deployment, providers and users require more sophisticated service architectures for effective Internet and corporate connectivity. In addition, the average user does not wish to see an ATM PVC or SVC service—they require end-to-end Internet connectivity, an IP-UNI.User-Network Interface

For this reason, the ATM VCCs as provisioned by the ADSL transport provider (the ILEC or CLEC) are really there only to support the end-to-end PPP and bridging services offered by their partner ISPs. Although bridged services predominated during the latter part of 1998 and into the beginning of 1999, these are beginning to transition to PPP, which will form the bulk of future deployments.

The advantages of relying on PPP in the DSL space are equivalent to those in the dial space. These include

- Per-session authentication based on PAP, CHAP, or token-card systems
- Optional Layer 3 address auto-configuration via DHCP
- Establishing multiple concurrent sessions (either via separate ATM VCCs or multiplexing via PPPoE; both options are described later in this chapter)
- Transparency to Layer 3 protocols (although IP is of course the most dominant)
- Per-session encryption and compression
- Per-session billing via interaction with RADIUS (or other) servers

Other architectures such as RFC-1483 bridging or even conventional routing from the CPE offer only a subset of these capabilities.

PPP was first standardized within the IETF in the 1989 time frame (RFC-1331), and provides a standard method of encapsulating higher-layer protocols across point-to-point connections such as leased lines and access links. Although based on the HDLC packet structure, it extends this structure with a 16 bit protocol identifier containing information as to the content of the packet. This packet may contain three types of information, as follows:

- The first is the Link Control Protocol (LCP), which negotiates session parameters such as the packet size or type of authentication.
- The second packet type consists of control frames for higher-layer protocols, used to determine that the receiver can in fact support the likes of IP, IPX, or DECnet. This is known as a Network Control Protocol (NCP), and in the case of IP, the protocol used is the IP Control Protocol (IPCP). These frames also negotiate any protocol-specific parameters. Note that each Layer 3 protocol has a specific control protocol used for these operations.
- Finally, the data frames contain the actual user data.

PPP over ATM (RFC-2364) builds upon the lessons learned in deploying PPP over frame-based networks. When applied to ATM, it does place some constraints upon the ATM network. Only AAL5 is applicable, and the PVCs or SVCs support only point-to-point operation (no multipoint connections are supported).

The CPE encapsulates the user data based on the VC-mux or LLC forms of RFC-1483, as depicted in Figure 4.5. In the majority of cases, the VC-mux encapsulation is preferable for a number of reasons. First, since the ATM VCC is dedicated to a single PPP session, the QoS capabilities of the VCC may align with the application requirements. Next, security for the PPP session will apply to all traffic over the VCC, eliminating the requirement for additional ATM-layer security. Finally, if the ATM VCC is used only to transport PPP, the additional LLC-SNAP overhead is not required. This reduces the amount of processing required at the endpoints of the connection. However, if the provider implements Frame Relay service interworking within the network (such as to communicate with frame-based CPE, both ADSL and non-ADSL), only LLC is acceptable.

If PPP over ATM is used in conjunction with SVCs, the Q.2931 B-LLI extension octet specifies an IPI value of 0xCF, signifying PPP. The service aggregator, if operating in PPP termination mode, strips the PPP over ATM header from the user's data and forwards it as routed traffic. Alternatively, the aggregator may rely on tunneling, MPLS, or IPSec for transport of the traffic to an upstream termination point.

Reusing PPP as deployed within Internet dial services, ATM enables a number of service models. These may be divided into two classes. PPP termination provides general access into an Internet core, whereas a different set of services based on policy routing (PTA, L2TP, IPSec, and even MPLS) offer VPN access to ISPs and corporate gateways.

FIGURE **4.5.**

PPP Architecture

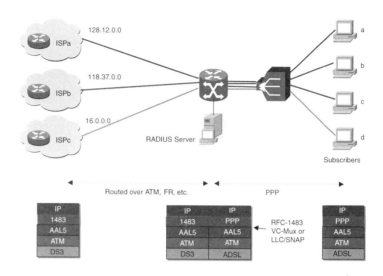

Note that none of these service models require SVCs, as they rely on Layer 3 mechanisms to establish the user sessions dynamically. Also relevant are the various methods of extending the PPP session across the subscriber's Ethernet segment to the PC(s). These include PPPoE, BMAP, and various forms of tunneling (reusing encapsulations such as L2TP and PPTP).

Relevant to both VPN and non-VPN architectures is the interworking of Layer 3 and ATM QoS. This is critical in enabling support for real-time applications and requires careful network design as well as the support of this interworking within the various network elements. For example, an ISP may define different categories of users, some with access to real-time services, such as VoIP, and some with access to best-effort only. When a user in the former category authenticates at the AAA server, this server could pass an attribute to the aggregator mapping the user's traffic to a higher IP precedence or into an ATM VCC with a higher QoS. This would give the subscriber in question a higher priority across the provider's backbone.

4.2.1 PPP Termination

Under PPP termination, the user PPP session extends from the CPE to the aggregator, where it terminates at the Layer 3 routing function. Therefore, all authentication takes place via the aggregator, either through an external AAA server or an internal configuration file.

With PPP termination, the ADSL transport provider will provision an ATM PVC from the user's CPE. This PVC extends across the ADSL network, terminating at the aggregator operated by a partner ISP or corporation.

At the subscriber's location, the termination point of the PVC may be at the PC itself or at an external ATU-R. If the termination point is a PC, it is equipped with an ADSL NIC. Alternatively, an ADSL modem will present an ATM25 interface to an ATM NIC, although this is less common. If the ATU-R presents a non-ATM connection to the PC over Ethernet or some other media, the PPP session extends to the PC. The section entitled "Extending PPP to the Desktop" details options for accomplishing this extension. Note that the QoS parameters of the PVC (such as bandwidth) need not equal those of the tariffed ADSL service, since traffic shaping is possible at the higher layers.

At the aggregator, the ADSL PVC terminates on a subinterface and is associated with a virtual template. These templates contain information as to the QoS of the PVC, detailed as part of the configuration examples in Chapter 5.

Provisioning is one area of concern when dealing with large numbers of PVCs. Rather than individually having to configure upwards of thousands of PVCs at the router, you can implement PVC discovery between the router and the last-hop ATM switch. Here, any PVCs configured automatically map to a router subinterface corresponding with their VPI. For example, all PVCs with VPI=10 map to subinterface 10.

Figure 4.6 shows the PPP termination events. Following this flow of events leading to session establishment within PPP termination and using Windows dial-up networking as an example, a user clicks on a desktop icon to initiate the PPP session, initiating LCP negotiation between the PC and the router (aggregator). The router now attempts authentication via CHAP, causing the password dialog window to appear on the desktop.

In the case of PPP termination, the user must enter only a username and password. The router now queries an external AAA server (such as RADIUS). If only a small number of subscribers is provisioned, this information may be stored local to the router instead.

FIGURE 4.6.

PPP Termination
Events

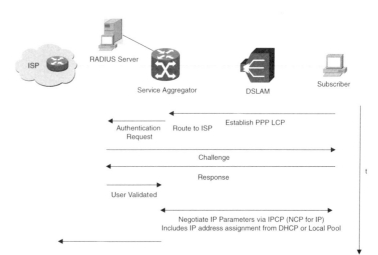

An option here is for the RADIUS server, upon user authentication, to return a set of attributes (including the IP address to be used in the IPCP phase) that could be used to modify the user's QoS parameters. As an example, an attribute may alter the PVC's QoS as defined in the virtual template.

4.2.2 Policy Routing

Now moving into the VPN space, the most basic way of implementing an IP-VPN is through the use of policy routing within the aggregator.

As with PPP termination, the ADSL provider preprovisions a PVC between the subscriber's CPE and the partner ISPs aggregator. Within the aggregator, route maps associate sets of users, based on their source IP address, to upstream IP networks. These individual networks may be bound to PVCs or subinterfaces within the aggregator.

A user, upon authentication, is dynamically associated with a given upstream destination based on its RADIUS attributes. Although this service model meets the needs of ISPs, other techniques such as multiple context support, L2TP, and MPLS are more scalable. Figure 4.7 depicts a RADIUS-based policy routing implementation, where different users are bound to different upstream destinations based on their RADIUS authentication.

4

FIGURE **4.7.**

Policy Routing

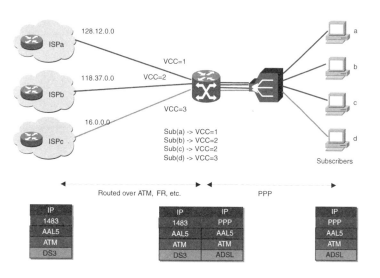

4.2.3 ISP Contexts

An alternative to the policy routing implementation described previously is one based on a service model supporting multiple independent forwarding (routing) tables, sometimes referred to as "contexts." This model is also known as PPP Terminating and Aggregation (PTA). Here, a user maps to an upstream interface pointing to a specific ISP or corporation.

One characteristic of this architecture is that the destination networks may have overlapping IP address spaces. For example, two corporations may both use addresses from the private network 10.0.0.0, and subscribers on each may have identical IP addresses. Ordinarily, a router would have problems dealing with these overlapping addresses. Support for multiple routing tables, in effect virtual routers, allows this to occur, with the IP address range of one context shielded from all others.

Following actual operation, the user initiates a PPP session terminating at the aggregator and enters the authentication phase. The user enters an ID in the format **user@domain**, with the domain portion of the login used to select the SVC or PVC at the upstream side of the aggregator.

The aggregator supports proxy RADIUS whereby it forwards the authentication request to the local RADIUS server, which then acts as a RADIUS client to servers operated by the destination ISPs or corporations. Figure 4.8 shows the complete forwarding process.

FIGURE **4.8.**

ISP Contexts

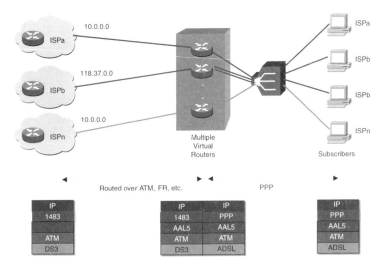

This local server sends the proxy RADIUS request containing the username, password, and challenge (if implemented) across the ISP or corporate VCC to the destination RADIUS server. Based on successful authentication, this server supplies an IP address, which is then forwarded to the aggregator and then to the subscriber during IPCP negotiation.

At the same time, the aggregator binds the subscriber to the destination VCC corresponding to the corporation or ISP. The subscriber's IP address is now part of the native IP address space of the destination, and all client traffic is routed to the destination. Here, no PPP session extends between the subscriber and the destination. Also, note that no per-subscriber configuration information is required on the RADIUS server associated with the aggregator. This is critical for scalability.

During data transfer, all traffic from the subscriber passes to the VCC assigned to the upstream provider. In the opposite (downstream) direction, the aggregator examines the packet's destination IP address and then forwards it to the subscriber associated with that IP address.

Each ISP or corporation has a unique forwarding table, populated from the RADIUS replies. A more complex situation results when the CPE is a router as opposed to a single PC. Here, the router requests a subnet of IP addresses via DHCP. This request is passed to the provider's RADIUS server by the aggregator using Proxy DHCP. Upon receipt of the subnet assignment, the aggregator updates its routing tables and then forwards the reply to the CPE.

4.2.4 L2TP – Layer Two Tunneling Protocol

Another method of constructing IP-VPNs is by using tunneling between the aggregator and a Home Gateway (HGW). This VPN architecture closely parallels the architecture used within dial, where the subscriber initiates a session to an NAS in the service provider's network. Based on the intended destination, be it ISP or corporate, the NAS initiates a tunnel to a HGW located at the ISP or corporate premises. All PPP traffic from this subscriber and any other subscribers accessing the same HGW from the same NAS flows over the same tunnel.

In the L2TP scenario, the user initiates a PPP session in the same way as described in the section on PPP termination. However, as opposed to the aggregator terminating the PPP session and authenticating locally, the aggregator directs the user's session into an L2TP tunnel for transport to the ISP or corporate destination.

Note that this UDP or GRE-based tunnel is not bound to ATM, and may operate over most WAN infrastructures. In addition, although tunneling is most closely associated with IP, the protocol supports IPX and NetBEUI as well.

L2TP may use PAP, CHAP, MS-CHAP, or token cards for authentication; IPSec, ECP, or CCP for encryption negotiation; and a choice of IPSec, RC4, DES, or Triple-DES for the actual encryption. Key management is based on the IKE. The protocol also supports compression, using either LZS or MPPC.

The aggregation device initiating the tunnel is known as an L2TP Access Concentrator (LAC) in L2TP terminology. It is operated by an ILEC, CLEC, or even an ISP offering Internet wholesale services. However, the first deployments are by ISPs, CLECs in the United States, and Access Network Providers elsewhere, although during the course of 1999 the ILECs and PTTs should offer these services as well.

Take a look at the L2TP data flow as shown in Figure 4.9. A PC user activates Windows dial-up networking and proceeds to login with a fully qualified username of the form **user@domain**. As with the PTA architecture, the domain portion identifies the intended destination. The LEC now parses the userid. It passes the request to a RADIUS server (as an example) for authentication. As part of this transaction, the server returns parameters directing the user's PPP session to a given L2TP tunnel identified by a VCI/VPI or E.164/NSAP address, and optionally associated with a specific QoS.

FIGURE 4.9.

L2TP Tunneling

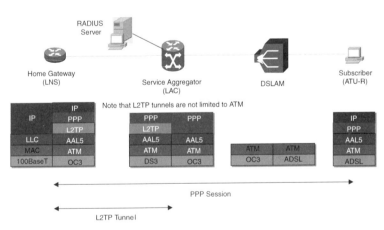

As depicted in Figure 4.9, a corporation may offer different classes of services, in this case gold and bronze. Users authenticating with **user@bronze.corp.com** are placed in a tunnel with bronze bandwidth and particular oversubscription factor, whereas other users will be placed in the gold tunnel offering higher performance. If the specified tunnel does not yet exist, the aggregator creates the tunnel in the following way:

1. First, the aggregator issues an "L2TP incoming call request" message to the L2TP Network Server (LNS) stating that it wishes to set up a subscriber PPP session.

2. The Home Gateway, or LNS, replies with an "L2TP incoming call reply" to acknowledge that it is willing to set up the session.

3. The LAC in turn responds with an "L2TP incoming call connected" message which contains the negotiated LCP parameters, the user's name, password, and challenge string (CHAP).

4. Finally, the LNS authenticates the subscriber via its local AAA server and assigns an address via DHCP in the case of IP. If this last phase is unsuccessful, the LNS sends an "L2TP call disconnect notify" message, which causes the PPP session to be terminated. An advantage of this architecture is that no per-subscriber information must be stored in the LAC.

Although the previous discussion focuses on L2TP, two other tunneling protocols are in use. The first, the Layer 2 Forwarding (L2F) protocol, is implemented by a number of internetworking vendors for dial VPNs but not supported on any clients. A more recent protocol is Microsoft's Point-to-Point Tunneling Protocol (PPTP), supported on Windows. As with L2TP, PPTP is multi-protocol, and includes provisions for encryption (although without support for IPSec), compression, and address assignment. L2TP, in fact, is a merger of L2F and PPTP.

4.2.5 Multiprotocol Label Switching (MPLS)

Although L2TP tunneling, and to a lesser extent, PTA, begin to address the provider VPN requirements, they still have constraints in terms of scalability, routing, and manageability.

Taking a step back, consider classical VPN architectures based on Layer 2 techniques such as point-to-point leased lines, Frame Relay DLCIs, and ATM VCCs. Traditionally, organizations deploy a hub-and-spoke topology due to circuit costs. If the organization is structured in such a way that all traffic flows back to a central point, this may not present a problem.

However, where more complex any-to-any data flows are involved, as is increasingly the case, this architecture is clearly suboptimal. The alternative is to create a full or partial mesh of circuits. In the former case, the number of connections required is on the order of N^2, where N is the number of sites. These point-to-point circuits effectively decouple the underlying Layer 2 topology from the Layer 3 routing protocols, resulting in scalability problems if fully meshed, and suboptimal routing if not. The goal, then, is somehow to model the VPN as an Internet, where customer sites connect to an IP cloud. One solution is Multiprotocol Label Switching (MPLS), building upon an earlier Cisco proprietary technique known as *Tag Switching*. Another solution based on the deployment of IPSec tunnels is detailed in the next section.

At each customer site, one or more Customer Edge (CE) routers peer with one or more Provider Edge (PE) routers located within the service provider's network. Figure 4.10 shows this MPLS-VPN architecture.

FIGURE **4.10.**

MPLS-VPN Architecture

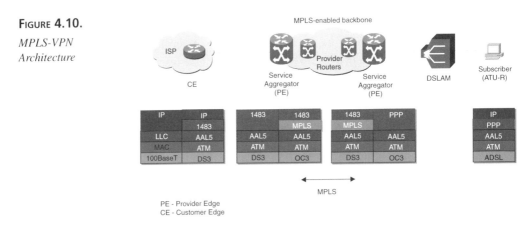

CE routers at different sites will not peer with one another, but will instead exchange routing information with the backbone via BGP or static routing. This requires some sophistication within the provider's network for the following reasons:

- First, customers with overlapping IP address spaces (such as those based on RFC-1918) must be able to pass their routing information across the backbone in a nonambiguous way.

- Next, the core routers in the backbone cannot be expected to hold routing information for every customer network.

- Finally, the provider must ensure that traffic from one customer's VPN has no way of reaching another customer's network. MPLS VPNs address the concerns. Each VPN is assigned a unique VPN-ID. PE routers, upon receipt of routing information from CE routers, will convert the subscriber's IPv4 addresses in use to VPN-IPv4 addresses, which consist of the VPN-ID prepended to the IPv4 address. PE routers connecting to the same VPN will exchange VPN-IPv4 routing information pertaining only to that particular VPN. This per-VPN routing information is hidden from the provider's core routers, which are responsible only for routing traffic between the PE routers.

To maintain separation of the routing tables, the PE routers maintain separate Forwarding Information Bases (FIBs) for each VPN. Now following the MPLS sequence of events when a packet arrives from a given CE router in a particular VPN (since a single CE router may in fact be a member of multiple VPNs), the following happens:

1. The PE router in question looks up the proper destination address in the corresponding per-VPN FIB and appends this label to the packet.
2. The ingress PE now uses its IGP routing table to determine the egress PE (or BGP next hop) for the packet and prepends a second label to the packet containing this routing information. This results in formation of a stack of labels.

Note that if the ingress and egress PE routers are directly adjacent via IP (such as Frame Relay or ATM connected), the first tag is not needed, whereas multiple connections between the PEs (one per VPN) would allow the second tag to be discarded as well.

Here, only the Multi-FIB would come into play. One other capability of MPLS-VPNs is the capability to separate traffic that originated within the VPN from traffic which did not. This allows the latter class of traffic to be directed to a firewall for further processing. Note that all PE routers must have MPLS functionality, although the CE routers do not. This is different than the L2TP model, which infers that every ISP or corporate home gateway is capable of tunnel termination.

To adapt MPLS-VPNs to the ADSL environment, what was previously a standalone Home Gateway at the ISP or corporate site now becomes a virtual entity located within a service provider's router. A user enters the network using dial-VPN techniques; it then transitions to MPLS-VPNs for transit across the provider's backbone. One advantage of this merging is that the MPLS CE and L2TP HGW functions may be combined into a single entity. Ultimately, the provider could further integrate these functions into a PE router to serve multiple VPNs.

There are in fact two MPLS-VPN deployment scenarios, depending on whether the operator of the aggregator is also acting as an ISP. In the first instance (shown in Figure 4.11), the two entities are separate, much like current L2TP deployments.

In the scenario shown in Figure 4.11, the aggregator operator establishes tunnels via L2TP to the ISP, which then forwards the traffic onto an MPLS backbone. During the subscriber login process, the ISP's AAA server returns the tunnel information.

FIGURE 4.11.

*MPLS–VPNs—Aggre-
gator as NAS, Separate
from Virtual Home
Gateway*

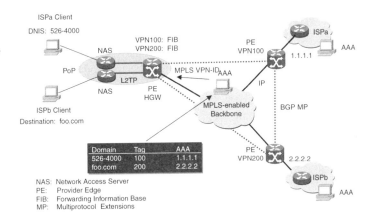

NAS: Network Access Server
PE: Provider Edge
FIB: Forwarding Information Base
MP: Multiprotocol Extensions

A simpler scenario, shown in Figure 4.12, is when a single organization performs both functions, eliminating the need for the tunnels. Here the virtual HG is located within the aggregator, and the AAA server need only return the VPN-ID.

FIGURE 4.12.

*MPLS–VPNs—Aggre-
gator as NAS and Vir-
tual Home Gateway*

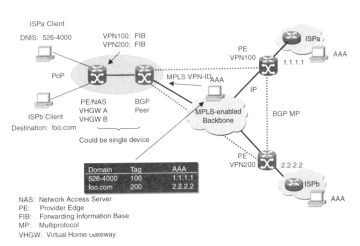

NAS: Network Access Server
PE: Provider Edge
FIB: Forwarding Information Base
MP: Multiprotocol
VHGW: Virtual Home Gateway

Note that this scenario allows two users of the service to communicate directly via the
aggregator, if policy permits, as opposed to traversing the ISP's backbone. There is a limitation, however, on the number of virtual HGWs supported by a single platform, a function of
the amount of per-VPN routing state required.

One option is to create the virtual HGWs on-demand based on active users. Here, only
active HGWs require routing information. Another concern is how to maintain a hub-and-
spoke architecture using MPLS–VPNs. This could be enabled by designating one PE as the
hub for a given VPN, with all other virtual HGWs belonging to that VPN forwarding traffic

to that point. Although routing may be less efficient than enabling any-to-any connectivity, the hub site is still capable of serving a much greater number of subscribers than that possible via L2TP tunneling since it must no longer maintain PPP state information.

> **NOTE**
>
> Note that the service provider should attempt to implement policy in such a way that routes all non-VPN traffic (traffic destined for the Internet) off the provider's backbone at the point closest to the user. A solution here is to create an Internet VPN with multiple exit points.

One common question asked is whether MPLS–VPNs support protocols other than IP. Although this is possible, it is not yet implemented, and GRE tunneling is the preferred approach.

Another common question relates to data encryption. If a user requires encryption of all traffic when traversing a shared backbone and if L2TP was used for part of the path, the virtual HGW will need to terminate the tunnel's IPSec when transitioning to MPLS–VPN encryption (at least based on current protocols). The scalability of this is still unknown.

Address assignment is also a concern and relates to the degree of routing optimization required. Every aggregator serving a particular VPN requires an address range for that VPN. One option is to assign subnets ahead of time, minimizing routing information at the expense of IP address usage. This is not a problem if using private addressing, but then Internet reachability becomes an issue. An alternative is to allocate the larger sites from pre-assigned ranges and to force the smaller locations to route through a smaller number of hub sites.

In the MPLS–VPN architecture, the flow of events is as follows:

1. The user first initiates a PPP session from the CPE, through the DSLAM, and into the aggregator acting both as an NAS/LAC as well as the PE/HGW.

2. At the conclusion of the PPP LCP negotiation phase with the NAS/LAC, authentication begins. Here, the aggregator identifies the user's intended VPN by parsing the fully qualified login string—**user@domain**— and then passing a "VPN Authorization Request" to the AAA server.

3. If the authorization request is successful, the server returns the VPN-ID, the BGP address of the VPN's border node, and possibly the DHCP server address and any QoS parameters.

4. The NAS/LAC now uses the VPN-ID to bind the subscriber's session to a per-VPN FIB. If this FIB exists, binding is complete; if not, it creates the new FIB and populates it with routing information via BGP (relying on the BGP node address returned).

5. The NAS/LAC may now inform the AAA server to begin accounting for this VPN session. PPP NCP now proceeds with DHCP address assignment.

6. Optionally, the aggregator may set the user's data to a certain IP Precedence based on the returned QoS parameter. End-to-end support of QoS across the MPLS backbone of course relies on a method of binding this QoS to a Label Switched Path (LSP), using either the Label Distribution Protocol (LDP) or RSVP.

4.2.6 IP Security (IPSec) Tunnels

Although a great deal of focus is on the creation of VPNs via L2TP or MPLS, VPNs may also be deployed based on the recently completed Internet security architecture, known as *IPSec*. IPSec is a self-contained set of standards providing for data integrity, confidentiality, and key exchange.

User authentication relies on a shared secret or a certification authority, while the Internet Key Exchange and the Internet Security Association and Key Management Protocol (IKE/ISAKMP) handle encryption negotiation as well as key management. DES, RC4, and Triple-DES are all valid encryption algorithms.

Depending on the level of security desired, a network administrator may decide to implement only the Authentication Header, insuring the integrity of the data. If the network backbone is unsecured, Encapsulating Security Payload enables encryption.

In both cases, the architecture includes a set of peer authentication, key creation, exchange procedures, and security policy negotiation collectively known as IKE. Finally, certificate management relies on the X509.V3 system for device authentication through a hierarchy of trusted sources.

As with L2TP, and in contrast to MPLS, IPSec VPNs are formed by establishing a set of secure point-to-point tunnels between sources and destinations. The standard is therefore very suitable for the typical DSL implementation in which multiple subscribers access an upstream ISP or corporate gateway.

Figure 4.13 depicts this IPSec architecture, with subscriber traffic encapsulated within the IPSec header for transport across an arbitrary backbone. Subscribers using RFC-1483 routed or bridged encapsulation may be mapped into the proper IPSec VPN via static configuration. Alternatively, if PPP encapsulated, the subscriber will first undergo authentication via a RADIUS server, the PPP session will be terminated, and the traffic will be forwarded into the VPN using 1483. Note that both of these options assume an ATM back-bone—nothing precludes the use of other technologies.

FIGURE 4.13.

IPSec Architecture

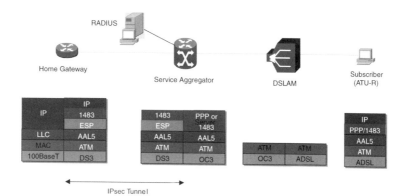

The downside of IPSec is its lack of support for individual user authentication and address management, two deficiencies addressed via external protocols such as RADIUS and DHCP.

4.2.7 Multidestination Support

One element that the previous service models share is that at any given point, the subscriber has secure access to a single upstream ISP or corporation. For example, under ISP Contexts the user's incoming connection binds to a single upstream VCC, and within L2TP tunneling, the PPP session extends end-to-end to a single HGW.

Extending all of these architectures and given proper intelligence within the network, a subscriber may in fact connect to multiple destinations at once. Here, a user has a connection to a primary destination, such as the Internet, and uses NAT for any secondary destinations (illustrated in Figure 4.14).

FIGURE 4.14.

Multidestination Support via NAT

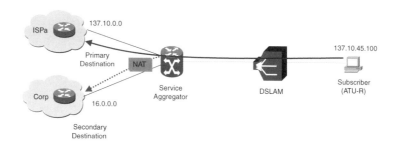

The primary destination is the one the user first logs into and authenticates, the ISP operating either the service aggregator or upstream ISP. In the former case, the local RADIUS server provides authentication, while the latter case invokes Proxy RADIUS to the upstream provider's server.

The RADIUS server in both cases now assigns the subscriber's IP address if it is a single user. If the subscriber has a router, Proxy DHCP provides the assignment. Note that this is identical to ISP Contexts. The difference here is that the service aggregator uses a special IP address for the control flow that comes into play when the subscriber wishes to access secondary destinations.

If the subscriber wishes to connect to a secondary destination (such as a corporate telecommuting service), the aggregator attempts to authenticate the subscriber with the destination in question, and optionally obtains an IP address depending on which actions are specified within the local RADIUS database.

Two methods for reaching this new destination are PPP tunneling via L2TP, or alternatively, RFC-1483 routing. In the former case, PPP authentication and address assignment via the IPCP takes place, while in the latter case, both operations rely on proxy RADIUS.

If this new secondary service provider implements address assignment (that is, they assign an address local to their address space to the subscriber), the aggregator implements NAT to map the subscriber's initial IP address to the new address assigned by the destination. This NAT allows, for example, the subscriber to appear native to a corporation's Intranet when telecommuting.

Here the service aggregator performs the NAT for all traffic flowing between the subscriber and the corporation, applying an IP address internal to the corporation on its upstream interface. Although there are some concerns with NAT regarding reachability, embedded IP addresses, and security, this service is still expected to see deployment due to its appeal.

Figure 4.15.

Multidestination Support via VCCs

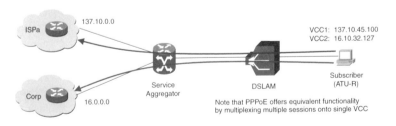

An alternative, highly secure method of providing connectivity to multiple destinations is through the use of multiple ATM VCCs. Consider a PC equipped with an ADSL or ATM25 NIC natively supporting multiple ATM circuits. Here, given support within the provider's

DSL service, one circuit could point at the public Internet while a second could terminate at the user's corporate gateway. Alternatively, PPPoE at the PC, described in the next section, allows the user to establish multiple PPP sessions terminating at more than one destination

4.2.8 Extending PPP to the Desktop

Although the service architectures described in the previous sections go a long way in enabling PPP connectivity across ADSL, one missing link has been the capability to extend the PPP session across the user's local Ethernet segment.

Revisiting the end-to-end deployment, a user will have a PC (or multiple PCs) connected to an ADSL modem, which then connects to the copper loop. If the ADSL modem is capable of terminating the PPP over ATM session, with PCs bridged across the local Ethernet segment, this is one solution.

Alternatively, some PCs may integrate ADSL modems, while an even smaller number will support ATM25 NICs connected to ADSL modems. However, to address the majority of the installations where the ADSL modem is capable only of bridging, but where the ADSL provider still wishes to migrate to the dial-centric PPP model, a method must exist to extend the PPP session easily to the PC.

One solution is PPP over Ethernet (PPPoE). PPPoE supports both single and multiple PC installations and is expected to become one of the predominate protocols in the consumer ADSL space.

An alternative, where ADSL modems implement the Broadband Modem Access Protocol (BMAP), is suitable for single PC installations but will be less common since it requires the BMAP capability within the ATU-R. In contrast, PPPoE requires no changes. Solutions based on implementing tunneling (L2TP, PPTP) from the PC to the aggregator are also expected to be less common due to complexity in the PC and required protocol support within the ATU-R. Within the ADSL Forum, the issue of Premises Distribution Network protocols has been a topic of much debate.

4.2.8.1 PPP over Ethernet (PPPoE)

PPPoE was proposed in 1998 as a way to PPP-enable the installed base of RFC-1483 bridge-mode ATU-Rs. It relies on support within the PC via a lightweight client (or shim), placing a small encapsulation header between the PPP header and the Ethernet MAC layer. Figure 4.16 shows the PPPoE protocol architecture.

FIGURE **4.16.**

*PPPoE Protocol
Architecture*

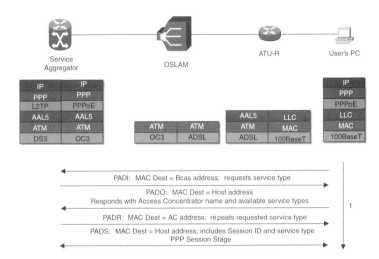

The service aggregator strips off this shim and processes the subscriber session as any other PPP over ATM session. This allows reuse of the various provisioning and AAA servers. One advantage of PPPoE is that it allows a PC to access multiple destinations simultaneously, as described next.

PPPoE operates as follows: The subscriber's CPE will first enter a discovery phase to identify the MAC address of the PPPoE peer (service aggregator) and to establish a session ID. It accomplishes this by sending out a PPPoE Active Discovery Initiation (PADI) packet to the Ethernet broadcast address. Multiple aggregators may respond by issuing PPPoE Active Discovery Offer (PADO) packets.

The CPE will then select one and respond with a PPPoE Active Discovery Request (PADR) packet to the chosen aggregator. Finally, this aggregator returns a PPPoE Active Discovery Session-confirmation (PADS) packet with a unique session ID. The session ID is the PPPoE multiplexing field within the PPPoE header (shown in Figure 4.17), which allows multiple sessions to traverse the same ATM VCC.

FIGURE 4.17.

PPPoE Header

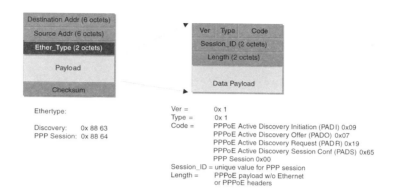

Ethertype:

Discovery: 0x 88 63
PPP Session: 0x 88 64

Ver =   0x 1
Type =   0x 1
Code =   PPPoE Active Discovery Initiation (PADI) 0x09
              PPPoE Active Discovery Offer (PADO) 0x07
              PPPoE Active Discovery Request (PADR) 0x19
              PPPoE Active Discovery Session Conf (PADS) 0x65
              PPP Session 0x00
Session_ID = unique value for PPP session
Length =   PPPoE payload w/o Ethernet
              or PPPoE headers

Upon receipt of the PADS, the CPE is now ready to enter the PPP session stage. Here it transmits PPP data as standard PPP data, with the ETHER_TYPE set to 0x8864 and the PPP protocol set to 0xCO21.

> One common question relating to PPPoE is whether it is immune to Denial of Service attacks. A solution is for the aggregator to limit the number of sessions.

4

PPPoE is capable of providing a PC with access to multiple destinations at a given time, although the mechanics of this are more a function of the operating system.

First consider a household with multiple PCs, each with a single PPPoE session open to a unique destination. Each PPPoE session is multiplexed over the in-house Ethernet, and then carried over the ATM VCC to the ADSL provider. This scenario is expected to suffice for the majority of the subscribers.

However, if a user wishes to open multiple sessions from the single PC, the applications must have some way of knowing what PPPoE session to use. For example, a spreadsheet binds to the session terminating at the corporation, while a browser may be used for Internet access.

The solution is to associate each PPPoE session with a different IP address and to equip the system with the intelligence to route as appropriate. If the corporate network assigns an IP of 16.10.154.13, and the address assigned by the subscriber's ISP is 137.10.32.100, any traffic destined to the corporation will be routed via the corporate PPPoE session. All other traffic will be routed to the ISP (shown in Figure 4.18), assuming that the PC has the necessary routing intelligence, available in Windows 98 and NT5.

FIGURE **4.18.**

*PPPoE Multi-
destination Support*

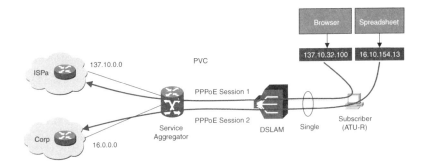

Note that the default Windows behavior is to route all traffic to the last PPP session established. Also note that even if the subscriber is capable of establishing two or more connections, this still may be contrary to corporate security policies.

4.2.8.2 Broadband Modem Access Protocol (BMAP)

An alternative to PPPoE for extending PPP to the PC is the Broadband Modem Access Protocol (BMAP). Unlike PPPoE, BMAP applies to USB and IEEE-1394 interfaces as well,
giving it added versatility in connecting newer PCs and laptops.

The protocol's downside is that it requires a BMAP-aware ATU-R, something that PPPoE
does not require. This drawback precludes the reuse of the installed base of bridge-mode
ADSL modems.

Using Ethernet as an example, BMAP data is carried as an Ethernet payload with a set
EtherType (see Figure 4.19). In the case of 802.3, it is LLC/SNAP encapsulated with
LLC=0x AA AA 03 and the SNAP header set to all zeros. BMAP, as currently implemented,
supports only a single PC communicating with an ATU-R at any one time.

FIGURE **4.19.**

BMAP Mode ATU-R

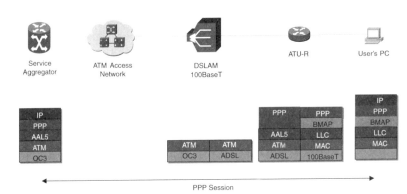

Although a form of BMAP supporting multiple PCs has been proposed, it is not expected to be implemented. This would rely on one PC assuming the role of a BMAP Server providing redirection for other BMAP clients on the LAN segment, whereby a single host will have access to a single PVC or SVC (ADSL, 1998a). Note that this implies one VCC per BMAP client, unlike PPPoE's support of multiple clients across a single VCC.

As with PPPoE, the PC follows a defined procedure for recognizing and then communicating with BMAP-enabled ADSL modems. At startup, it is in the DISCONNECT state. The PC now sends a BMAP DISCOVERY packet containing its supported data formats to the Ethernet broadcast address.

Note that USB or IEEE-1394 would allow the use of a device specific unicast address instead. The modem responds with a DISCOVERY_ACK containing its supported formats in order of preference (described shortly).

If multiple PCs on a LAN segment attempt to access the modem, the protocol establishes a binding with only one unless implementing a form of BMAP supporting multiple PCs. The PC and modem now enter the CONNECT state and begin to exchange data.

One characteristic of BMAP is its support of power managed PCs, allowing the modem and PC to maintain the binding even when the PC is powered down. It accomplishes this by defining a SLEEP message that tells the modem to maintain the binding even in the absence of POLL messages, which an active PC will periodically transmit.

BMAP supports three types of encapsulations across the LAN segment, depending on where the SAR takes place:

- Under Type 1, the PC transmits a number of 53 byte ATM cells encapsulated in an Ethernet frame. The PC performs all SAR functions and traffic shaping, reducing the cost of the ADSL modem at the expense of PC complexity. Type 1 supports all AAL types, along with ATM OAM and RM flows. The latter is required for ABR service.

- Type 2 removes duplicate cell headers, with the PC still performing the SAR. Here it transmits only one copy of the cell header at the beginning of the Ethernet frame. As with Type 1, it supports all AALs. Both types provide support for real-time data by allowing a frame to be sent with one or a few number of cells (AAL2, for example).

- In contrast, Type 3, with the SAR performed at the ADSL modem, supports only AAL5. Here, the PC generates the AA5 SDU prepended with a chosen VPI/VCI combination. Type 3 does not support OAM or RM cells.

4

4.2.9 L2TP to the Desktop

Alternative solutions include those based on creating an L2TP or PPTP tunnel between the user's PC and the ATU-R. This ATU-R could be a standalone router or a suitably equipped server. The user's PC encapsulates the TCP/IP traffic within PPP, passing it into L2TP or PPTP much like the role that the aggregator plays in the tunneling architecture.

However, this use of tunneling is not to enable VPNs per se; it is only to provide for PPP transport across the local Ethernet segment. The ATU-R then terminates the L2TP or PPTP tunnel, passing the higher-layer traffic across the ADSL loop and to the aggregator or home gateway where the PPP session terminates. The entire PPTP (or L2TP) to the desktop process is illustrated in Figure 4.20.

FIGURE 4.20.

L2TP or PPTP to the Desktop

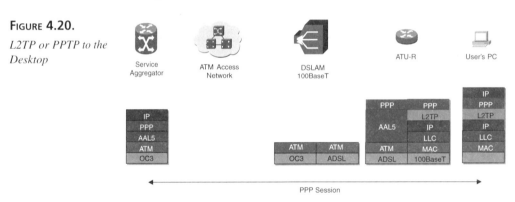

When a user wishes to initiate a PPP session, the PC establishes an L2TP tunnel to the ATU-R by exchanging L2TP control messages. As with other PPP architectures, the PC receives its address assignment via DHCP.

The ATU-R, upon receipt of the session request, will either pass it into an existing PVC to the proper destination or, if SVC-enabled, will signal a new connection, identifying it as a PPP session (based on the BLLI field). For this to occur, the L2TP request must either contain an IP destination that may be mapped to an ATM NSAP or E.164 address, or alternatively, the request may contain the ATM destination address. Once the ATU-R has established this connection, the PPP session will run end-to-end.

Although this L2TP to the desktop architecture has some merit, it does introduce additional encapsulation complexity across the LAN segment and, more important, requires an L2TP-capable ATU-R. Since one of the major criteria in PPP enabling the PC is the reuse of existing, deployed ADSL modems, alternatives such as PPPoE have greater appeal and will likely see wider deployment.

4.2.10 PPP Proxy

The options outlined in the three previous sections all focus on extending PPP over the LAN segment. A different solution terminates the PPP session at the ATU-R, while still providing the PC with functionality equivalent to PPP.

Note that this requires deployment of a suitably equipped router or server acting as the ATU-R. Under PPP proxy, the subscriber's ATU-R appears as a PPP host to the ADSL provider. Hosts behind the ATU-R may have different IP addresses or may share a single globally reachable address.

When the hosts have different IP addresses, each host will rely on a separate ATM VCC from the ATU-R to the ADSL provider. When the hosts share addresses, the ATU-R implements NAT, with local IP addresses translated to the globally reachable IP address of the PPP session (see Figure 4.21). Hosts receive their local IP addresses via static configuration or DHCP (given this capability on the ATU-R).

FIGURE 4.21.

PPP Proxy

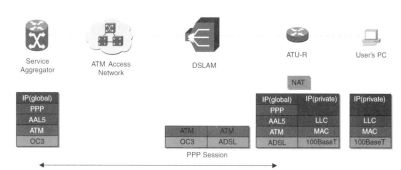

4.3 Bridging

A completely different way of enabling subscriber connectivity relies on the reuse of RFC-1483 bridging between the subscriber and the ADSL provider's PoP. This bridging architecture was in fact the first means of ADSL connectivity deployed, due to limitations within the first generation of ATU-Rs which were capable only of bridging, coupled with a lack of scalable PPP support at the PoP.

It is expected that bridged access will all but disappear in the next few yesrs, being replaced by options such as PPPoE due to its greater manageability, accountability, and security. The security issue is probably the most important consideration of all.

Bridging is a technology inherently suitable for trusted or closely administered environments, such as telecommuter or corporate remote site connectivity. Most bridges simply learn the location of a particular station by listening to the traffic on a port or VC and "gleaning" or obtaining source MAC addresses from the traffic.

Thereafter, traffic directed to the MAC address is sent to the port from which the address was gleaned. This type of bridging learning enables plug and play, but it is also a security hole since it opens up the possibility of MAC address spoofing and connection hijacking.

Another source of problems lies with the use of ARP to direct layer 3 traffic. An ARP request is simply a "where are you" query sent from a router to a station. The corresponding ARP reply is a "here I am" indicating to which MAC address (and hence which VC) to send the traffic for the layer 3 address. As one can imagine, ARP replies from two different VCs, whether due to malicious intent or misconfiguration, can have a drastic effect on routing.

In the first bridged deployments, the ATU-R performed simple interworking between the Ethernet MAC and the RFC-1483 encapsulation used across the ADSL loop. At the PoP, a router or even a LAN switch terminated the bridged traffic, routing it into an Internet backbone or, alternatively, extended the bridged traffic itself to an upstream destination (no routing).

More sophisticated architectures are based on the introduction of a function known as *subscriber bridging* and by deploying bridge groups for additional security. An even more secure architecture is based on what is known as *half bridging*.

Note that bridging does not preclude mult-destination support. In fact, this capability is implemented by at least two vendors. In addition, the concept of contexts within the aggregator, which allows for multiple independent routing tables, holds as well.

4.3.1 RFC-1483

In the ADSL space, bridging relies on the reuse of the RFC-1483 encapsulation originally defined for transporting bridged PDUs across native ATM. Although RFC-1483 bridging saw some use in campus deployments and in some MANs, it was all but relegated to the sideline until the advent of ADSL, with techniques such as RFC-1483/1577 routing, LANE, and MPOA enjoying much greater deployment. Figure 4.22 depicts both the LLC/SNAP and VC-Mux forms of RFC-1483.

Under RFC-1483 bridging as defined in Heinanen, (1993), an Ethernet-connected PC encapsulates the IP (or non-IP) traffic in an Ethernet frame and forwards it to the ATU-R, which encapsulates the Ethernet frame within RFC-1483.

FIGURE **4.22.**

RFC-1483 Bridged Encapsulations

FIGURE **4.22.**

RFC-1483 Bridged Encapsulations

The data then passes through the SAR before being transmitted across the ADSL loop. It passes transparently through the DSLAM, finally arriving at a router or LAN switch (Figure 4.23).

FIGURE **4.23.**

RFC-1483 Protocol Architecture

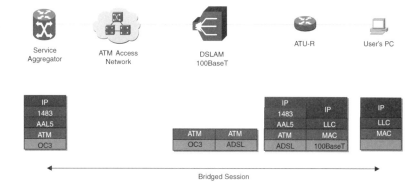

If the data arrives at a router, it passes into Layer 3 via an IRB-like function, or alternatively, the router swaps the RFC-1483 encapsulation for another bridged encapsulation (such as for Frame Relay IAW RFC-2427) before forwarding it on to its ultimate destination.

This architecture is preferred by some users who wish to emulate a LAN segment across ADSL. If a LAN switch is deployed instead of a router, an end-to-end bridged encapsulation is all that is available. Note that some degree of security is still possible under this architecture if the router or LAN switch is capable of handling VLANs to segment users or groups of users.

Popular techniques include Cisco's ISL and IEEE 802.1q. In both cases, traffic from different groups of users is placed into separate VLANs, allowing logical separation within the PoP.

4.3.2 Subscriber Bridging and Bridge Groups

Additional security and manageability for bridged subscribers are enabled by deploying a combination of subscriber bridging and bridge groups within the PoP. These techniques begin to introduce Layer 3 functionality into what was traditionally a Layer 2 service.

Consider an RFC-1483 deployment, with subscriber VCCs arriving at the aggregator on a single subinterface. They all share the same IP subnet and, due to the nature of bridging, are all capable of receiving each other's broadcast and multicast traffic (although not unicast).

An additional problem is that the aggregator determines which downstream VCC receives which traffic via ARP broadcasts and MAC address learning, thus making it relatively easy for a malicious user to hijack another's traffic.

Subscriber bridging addresses the first concern by allowing the network administrator selectively to configure filters between users. For example, a corporation or university may want to emulate a LAN and therefore would want leakage of most traffic. In this case, the provider sets no filters (as shown in Figure 4.24).

FIGURE **4.24.**

Subscriber Bridging

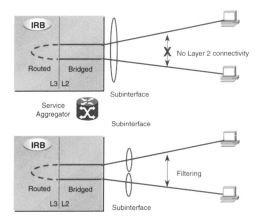

In contrast, the average telecommuter or corporate stub site has no reason to receive the broadcast traffic intended for another. Here the provider implements filters between each and every subscriber VCC. This filtering also requires that all traffic from one subscriber to another flow through an ISP's router, which is desirable in any case.

Closely related to subscriber bridging is the capability to configure multiple bridge groups. These are separate Layer 3 domains allowing the service provider to segment groups of users based on policy. For example, an aggregator may serve subscribers belonging to a number of ISPs or corporations.

With bridge groups (as shown in Figure 4-25), each ISP and corporation is assigned a separate IP subnet, with all traffic routed from one subnet to another.

FIGURE **4.25.**

Bridge Groups

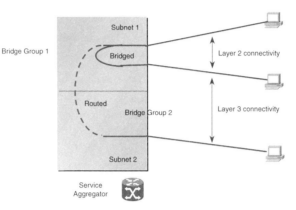

Within a bridge group, the provider may still deploy filters as described previously. Subscriber bridging improves scalability but is still vulnerable to MAC spoofing and ARP hijacking. The following section describes the solution for this form of attack—called *half bridging*.

4.3.3 Half Bridging

Half bridging provides a solution to spoofing and traffic hijacking by changing the way in which the aggregator processes subscriber packets. As opposed to passing the packets to the Layer 2 engine within the router with no control, half bridging allows the service provider to control which Layer 3 address goes over which VC and does not rely on ARPs to do this resolution, as shown in Figure 4.26.

FIGURE **4.26.**

Half Bridging

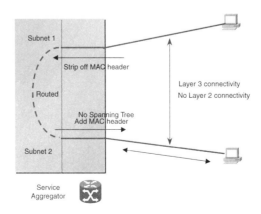

The subscriber, however, implements only bridging, thus the term *half bridging.* Each subscriber is treated as an individual Layer 3 entity in terms of address assignment and security at the aggregator. Even if a given subscriber were to spoof his or her MAC address, all decisions at the aggregator are made at the IP Layer so the MAC address is irrelevant. Note that most service aggregation platforms handle bridged subscribers in this way, even though they may not refer to the technique as half-bridging.

Next, if the same subscriber were to attempt to spoof or misconfigure his or her IP address, it would appear that the aggregator was an invalid address, not corresponding to the address set at the subscriber's interface on the ISP router. The data would then be discarded. Incoming data is always delivered to the proper subscriber based on the destination IP address configured by the service provider.

An additional feature of half bridging is the elimination of the spanning tree protocol across the subscriber loop, resulting in less processing required at the aggregator. This elimination is not really a problem, since subscribers are assumed to be singly homed to the service (at least at Layer 2). Single homing eliminates the need to manage primary and backup bridged paths.

4.4 Routing

Unlike PPP-based access models that emulate dial service and bridging that is more-or-less an interim solution, a fully routed service provides the equivalent of leased-line or Frame Relay connectivity across ADSL.

For example, a small business may currently access the Internet via a dedicated T1 link by running a dynamic routing protocol between its router and that of its ISP and relying on an ISDN connection for backup. To transition to ADSL, the provider must offer the same services. This is the role of RFC-1483 routing (shown in Figure 4.27), with deployment expected by many businesses connecting to ADSL.

FIGURE 4.27.

RFC-1483 Routing

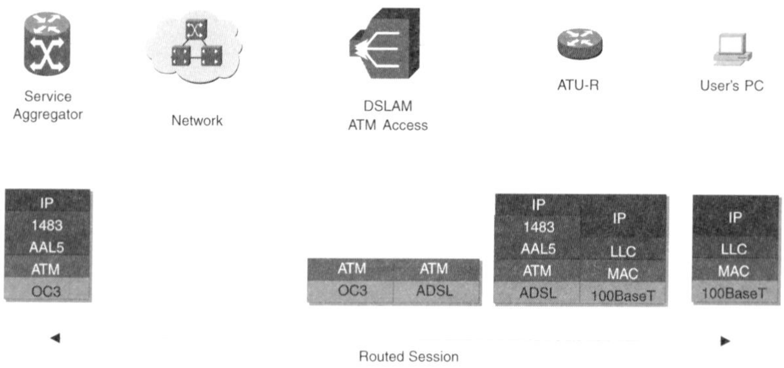

Service Aggregator   Network   DSLAM ATM Access   ATU-R   User's PC

Routed Session

Although these businesses initially rely on PVCs, nothing precludes the deployment of SVCs in conjunction with an RFC-1577 based ATMARP Server, given support within the provider's DSLAMs and ATM network. Aggregation vendors predict that up to 10 percent of ADSL subscribers will require this routed service.

Note that this RFC-1483 encapsulation is no different than that used across any other ATM access link except for the use of ADSL at the physical layer.

4.5 Voice

Parallel to the various data service models is the requirement to support voice across the ADSL infrastructure, an area of focus within the ADSL Forum. There are many ways to support this service depending on the business model of the service provider, whether the customer is residential or business, and the type of CPE and infrastructure deployed.

Options include analog voice splitters (as described in Chapters 2 and 3), Voice over ATM (VTOA), and Voice over IP (VoIP). All three of these options exist, and in fact, a service offering may consist of more than one. In addition, even deployment of voice splitters across the ADSL loop does not preclude interworking to VTOA or VoIP within the central office.

Splitters are currently the most common method for delivering voice to the subscriber, and ADSL's capability to support the traditional telephony traffic in the baseband is a major selling point.

In most cases, this voice traffic enters the MDF within the Central Office and is then shunted to the existing PSTN switch (such as 5ESS). Alternatively, the provider may deploy an ATM or IP-centric core infrastructure which eliminates the need for the switch. In this case, a suitable interworking device would convert the analog (or ISDN) voice traffic to either ATM or IP for transport across the backbone.

In the case of ATM, the required capabilities exist today, whereas necessary intelligence within IP devices (such as MGCP and SS7) should be widely available in the next year. Figure 4.28 depicts the splitter architecture.

FIGURE **4.28.**

Voice-Splitter

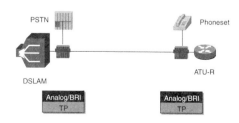

The next option supports voice in-band, within the ADSL data stream using VTOA or VoIP. In the case of VTOA, suitably equipped ADSL CPE converts the analog voice traffic to AAL1 or AAL2 for transport across the ADSL loop (Figure 4.29). Of course, the CPE (ATU-R) must include the necessary queuing to provide the necessary QoS support to the voice traffic.

FIGURE 4.29.

Voice-ATM

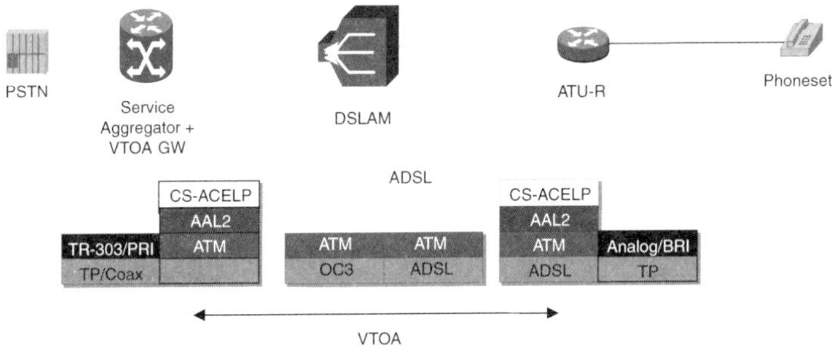

The DSLAM and aggregator must also support ATM QoS. At the aggregator, the VCCs carrying the voice traffic are either trunked to an external interworking unit or, alternatively, converted to traditional T1/E1 voice trunks for interworking with the PSTN. Another option is to convert the VTOA to VoIP within the aggregator.

Probably the most interesting architecture is based on VoIP. Here, the CPE converts the analog voice traffic to VoIP, forwarding it over the ADSL loop to the aggregator. Just as with VTOA, the CPE and DSLAM must provide the required QoS support. The most likely solution is to carry the VoIP traffic over a separate VCC, buffered separately in the ATU-R and within the DSLAM.

At the aggregator, the VoIP traffic is either forwarded into a VoIP backbone or converted to T1/E1 trunking for PSTN interworking. In either case, the provider's network must support the necessary call management and intelligence. This is the role of MGCP within the interworking unit, along with the support of SS7 for call signaling. Figure 4.30 depicts an end-to-end VoIP service.

Taking voice integration one step further, some vendors have proposed a next-generation DSLAM architecture in which both the ADSL data and analog voice terminate on the same linecards. Upstream interfaces will include an ATM trunk for the data and one or more T1/E1 links (TR-303, ISDN-PRI, and so on) for the voice traffic.

Additional versatility may be enabled by providing the linecards with on-board VoIP conversion, although this would imply Layer 3 functionality in the DSLAM. Note that one potential problem with this tight integration of the data and voice traffic is the state of

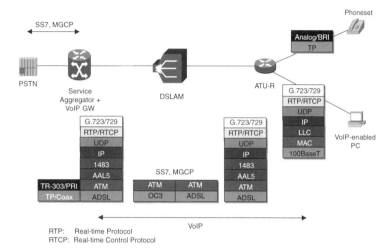

FIGURE 4.30.

Voice-End-to-End VoIP

RTP: Real-time Protocol
RTCP: Real-time Control Protocol

regulation in the United States regarding which providers may offer what types of services. For example, one eventuality may be that DSL services will be operated by CLEC components of the RBOCs, precluded from offering voice services. This would preclude analog voice or VoIP interworking.

4.6 Video

The transport of video across the ADSL loop, either IP-based or encapsulating MPEG-2 directly over ATM, is as interesting a service possibility as the transport of voice. The former IP-based solution is really a special case of any of the data transport models described earlier, but it does impose different QoS requirements upon the network (in the same way as VoIP demands some QoS guarantees).

Looking first at IP video, a server unicasts or multicasts MPEG-1 and in some cases MPEG-2 video. If this video over IP is a special service, the provider provisions a separate VCC from the content source or aggregator to the subscriber (as shown in Figure 4.30).

The DSLAM and ATU-R along the path now provide this VCC with the necessary QoS to support the video stream. Note that the video server need not be ATM connected. If attached to FE, for example, the aggregator will only need to know which set of subscriber VCCs are to receive the signal.

Here, the aggregator supports IP multicasting, with only a few video streams flowing over the FE between the video server and the aggregator. The number of discrete streams would depend on whether the service is prescheduled, Near-Video-on-Demand (NvoD) where specific content is periodically screened (i.e., every 15 minutes), or true VoD.

In contrast to IP-based video, the broadcast industry prefers the direct encapsulation of MPEG-2 over ATM, following MPEG-2s adoption within the industry as the encoder of choice for digital television and DVD.

In this scenario, the video server is ATM-attached, with a unique PVC spanning from the server to the PC or STB serving each subscriber. Later, SVCs will come into play, thus avoiding the need to preprovision subscriber connections. An additional optimization of this ATM video service relies on a single pt-mpt VCC for video distribution to multiple subscribers. This of course requires additional control within the network.

As with IP video, the aggregator, DSLAM, and CPE must properly prioritize the MPEG-2 VCCs in relation to any other best-effort VCCs in use. Figure 4.31 depicts this architecture.

FIGURE **4.31.**

Video - ATM

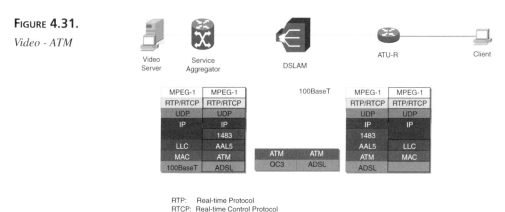

RTP: Real-time Protocol
RTCP: Real-time Control Protocol

4.7 Portals

Although the end-to-end service models described in the previous sections provide the basis for delivering subscriber services, there is another element within the network which may go a long way in providing users with better access and visibility to these services.

This is the concept of a network "portal," a user's window into the services and destinations offered by the service provider. It permits an ISP to differentiate services, by offering a customized experience on a per subscriber basis. In addition, if the subscriber is always directed to this portal upon authentication, a "mandatory portal," the ISP has a greater lock on the user. It enables "sticky services" which follow the subscriber and which help to build loyalty and therefore minimize churn. Figure 4.32 depicts a portal implementation, showing the interaction between the subscriber, the service aggregator, the RADIUS server, and the portal server.

FIGURE 4.32.

Portals

Notes:

IP-based control plane is not depicted
MPEG2 in IP encapsulation also possible across AAL5

Summary

In this chapter, the various ADSL service models were introduced. These include end-to-end ATM architectures based on PVCs or SVCs, PPP from the subscriber to the network provider (including extending PPP to the desktop), bridging models, fully routed access, and integration of both voice and video as part of a service architecture. Finally, a recent, much publicized part of service delivery known as portals was introduced. These service models, in combination with an understanding of the technology and protocols described in Chapters 2 and 3, provide a basis for end-to-end service deployment. Chapter 5 combines everything described thus far as part of these service rollouts.

Endnotes

Brown, C. and Malis, A., "RFC 2427: Multiprotocol Interconnect over Frame Relay," IETF, September 1998.

Gross, G., Kaycee, M., Li, A., Malis, A., and Stephens, J., "RFC 2364: PPP over AAL5," IETF, July 1998.

Harkins, D. and Carrel, D., "RFC 2409: The Internet Key Exchange (IKE)," IETF, November 1998.

Kent, S. and Atkinson, R., Security Architecture for the Internet Protocol, draft-ietf-ipsec-arch-sec-07.txt, July 1998.

Kent, S. and Atkinson, R., "RFC 2402: IP Authentication Header,", IETF, November 1998.

Kent, S. and Atkinson, R., "RFC 2406: IP Encapsulating Security Payload (ESP)," IETF, November 1998.

Kwok, T., et al., "An Interoperable End-to-end Broadband Service Architecture over ADSL," ADSL Forum, December 1997.

Mamakos, L., et al., PPP over Ethernet "PPPoE," Internet Draft, draft-carrel-info-pppoe-01.txt, September 1998.

Maughan, D., Schertler, M., Schneider, M., and Turner, J., "RFC 2408: Internet Security Association and Key Management Protocol (ISAKMP)," IETF, November 1998.

Tai, C., et al, "BMAP: Extending PPP/ATM Services across Ethernet/USB/IEEE 1394," ADSL Forum 98-018, March 1998.

Tai, C., et al, "Multihost Support for BMAP," ADSL Forum 98-019, March 1998.

Chapter 5

ADSL Implementation Examples

This chapter builds on the underlying ADSL technologies described in Chapter 2, the protocols introduced in Chapter 3, and the service architectures outlined in Chapter 4 by presenting a number of complete configuration examples. These examples are presented as applications and services visible to the subscriber, roughly following the ordering in Chapter 4. They include configurations utilizing systems from Shasta Networks, 3Com, and Cisco Systems.

The examples include the ones listed in the following table:

Application	Subscriber Encapsulation	ILEC	ISP/Corporation
1 Internet Access	RFC-1483 bridged, routed, and PPP over ATM	ATM VCCs	Terminates PVCs and routes across trunks
2 The Last Hundred Meters	PPP over Ethernet (PPPoE)	ATM VCCs	Terminates PPPoE
3 Beyond Aggregation	RFC-1483 bridged, routed, and PPP over ATM	ATM VCCs	Terminates PVCs and applies services
4 Internet Wholesaling	Any	SSG to multiple ISPs	Terminates routed sessions
5 Portals	Any	Operates Portal Server	Operates Portal Server
6 Corporate Intranet Access: VPDN	PPP over ATM	Tunnels PPP into L2TP	Terminates tunnels
7 Corporate Intranet Connectivity: VPRN	RFC-1483 routed	ATM VCCs	Tunnels subscriber sessions via IPSec
8 PSTN Bypass: VoIP	RFC-1483 routed and PPP over ATM	ATM VCCs	Terminates PVCs and routes to trunks and PSTN gateway
9 Entertainment: Video Streaming	IP-based MPEG-2 over ATM	ATM VCCs to video server	NA
10 Performance Testing	PPP over ATM	NA	NA

Before delving into the actual implementation examples, a brief review of the encapsulations found across the ADSL link is in order. These include RFC-1483 routed and bridged, as well as PPP over ATM. In addition, L2TP is commonly used for secure tunneling.

The RFC-1483 routed encapsulation is commonly used for small business access as well as for trunk connections. RFC-1483 Routed LLC permits multiple higher-layer protocols to share the same VCC, while the VCMUX encapsulation dedicates the connection to a single protocol. This does not preclude support of L2TP tunneling which will support multiple PPP sessions.

```
1483 Routed LLC:       AA-AA-03-00-00-00-XX-XX-[L3 packet]-[AAL5 trailer]
1483 Routed VCMUX:     [L3 packet]-[AAL5 trailer]
```

The RFC-1483 bridged encapsulation is used for single subscriber access. Over time, PPP (and PPPoE) will replace this encapsulation in most deployments.

```
1483 Bridged LLC: AA-AA-03-00-80-C2-00-07-[00-00]-[L2 frame]-[AAL5 trailer]
       (00-07 is 802.3 with no FCS)
1483 Bridged VCMUX: [00-00]-[L2 frame]-[AAL5 trailer]
```

PPP over ATM, based on RFC-2364, applies the dial access model to ADSL. As with RFC-1483, both LLL and VCMUX encapsulations are possible. The majority of current deployments are based on LLC.

```
PPP LLC:       FE-FE-03-CF-[00-21]-[IP packet]-[AAL5 trailer]
PPP VCMUX:     [00-21]-[IP packet]-[AAL5 trailer]
```

Across the trunk, L2TP will commonly support secure tunneling of multiple PPP sessions.

```
L2TP/AAL5:     [L2TP header]-[L2TP payload]-[AAL5 trailer]
L2TP header:   [flags][length]-[tunnel id]-[call id]-[NS]-[NR]-[offset
                   size]-[offset pad]
L2TP payload:  [00-21]-[IP packet]
```

The following sections now examine each example in detail.

5.1 Internet Access: Residential and Corporate

Scenario: An ISP, Shastanets.net, wishes to deliver Internet connectivity via an ADSL service operated by the local ILEC. This will allow the ISP to expand its service offerings beyond analog modems and ISDN, allowing it to compete better with the cable providers.

Solution: Shastanets.net partners with the ILEC delivering the ADSL service to offer Internet services to both residential and business subscribers. Although the type of CPE and supported data encapsulations are dependent upon the DSLAM deployed, in this case RFC-1483 bridging and routing as well as PPP over ATM are all supported as subscriber access encapsulations.

This permits Shastanets to meet the needs of its different customers better. The ILEC delivers the subscriber PVCs via an ATM trunk. The ISP terminates this link on a Shasta Networks Service Selection Gateway (SSG-5000), a service aggregator, and routes all subscriber traffic into its backbone. Terminating all subscriber PVCs at the aggregator of course minimizes the number of VCCs traversing Shasta's backbone.

Topology: As shown in Figure 5.1, the ILEC deploys the DSLAMs within its Central Offices along with the ADSL CPE. Remember that the choice of CPE is dependent upon the choice of the DSLAM. The DSLAMs in turn connect to an ATM access network, which allows the ILEC to combine the PVCs from the DSLAMs into a single trunk terminating at the ISP. This trunk connects to Shastanets.net's service aggregator, located at the ISP's PoP.

5

One interface on this aggregator connects to the ADSL provider, while two others connect to the ISP's backbone. Residential and business subscribers connect to the ILEC's ADSL service, with their PVCs forwarded to Shastanets' aggregator.

FIGURE 5.1.

Aggregation Topology

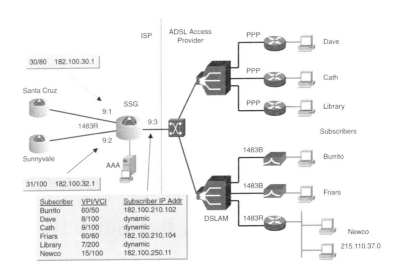

> **NOTE**
>
> Note that the ADSL access architecture described here applies to every example. It is therefore only detailed as part of this initial example.

Here, the ILEC actually deploys two types of DSLAMs, one supporting bridge-mode ATU-Rs and the other supporting routing.

Small businesses connect to the bridge-mode ATU-Rs via external routers. In this example, the CLEC has deployed the following equipment:

Subscriber Type	Encapsulation	DSLAM	CPE
Residential-PPP	PPP over ATM	Cisco 6100	Cisco 677
Small Business	RFC-1483 Bridged	Alcatel ASAM	Alcatel SpeedTouch Home
Small Business	RFC-1483 Routed	Alcatel ASAM	3Com 811

5.1.1 Review of DSL Deployment

The ILEC or CLEC will deploy the DSLAM and CPE and will hand off subscriber PVCs to the ISP across an ATM trunk. In most cases, multiple DSLAMs located at the Central Offices will feed into a regional ATM network, with the ISP trunk at the upstream side of this network.

Subscribers will normally contract with an ISP for DSL-based connectivity. All financial dealings will be between the subscriber and the ISP, with the subscriber having no direct contact with the ILEC or CLEC actually installing the service. The ISP, in turn, will contract with the ILEC or CLEC for CPE installation and loop bringup.

The ILEC or CLEC will perform the basic loop qualification and verify connectivity of the CPE. The ISP then takes over for final CPE configuration (if any is required, noting that some bridge-mode ATU-Rs and PPP-mode routers will require no customer-specific configuration).

In some instances, two CPE devices are installed. The first is the bridge-mode ATU-R deployed by the ILEC/CLEC, providing basic loop termination. In some European countries, this analogy to the ISDN NT1 will be common. The ISP will then install a more sophisticated router, totally under its control. This router will connect to the ATU-R via Fast Ethernet, ATM25, or USB.

The ISP also terminates the subscriber PVCs at a service aggregator, where it feeds the routed traffic into its IP backbone. In this configuration, the ILEC/CLEC provides transport only and has no view into the end-to-end Layer 3 configuration. Here, the ISP has full control over the router. A hybrid situation results when the ATU-R integrates routing functionality. Here, the ILEC/CLEC may conduct basic line configuration and then hand off control of the router to the ISP for further configuration. Figure 5.2 details these options.

If the ILEC/CLEC is wholesaling DSL services, the situation is a bit different. Here, the DSL transport provider terminates the subscriber PVCs at the service aggregator, handing off subscriber sessions to ISP clients in one of two ways. The first is via L2TP tunneling, while the second relies on multiple routing "contexts" within the platform. These options are described in sections 5.6 and 5.4.

5

FIGURE **5.2.**

Deployment Options

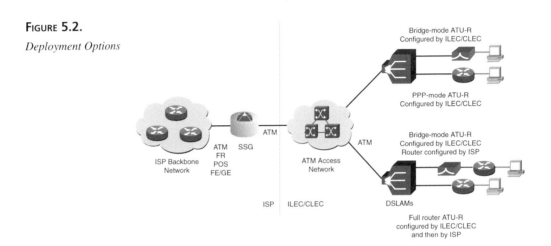

5.1.2 ILEC Preparation

The ILEC must first conduct site surveys at its Central Offices, as well as qualify the local loops for ADSL service. In the former case, it insures that adequate space, power, and cooling exists within each CO targeted for ADSL service. It also plans for the necessary ATM trunks to connect the DSLAMs to its ATM access network and for trunks to connect to partner ISPs (such as Shastanets.net). Upon DSLAM installation, it will establish the necessary PVCs between the subscribers, DSLAMs, and ATM access network via its existing provisioning system along with any vendor-specific DSLAM provisioning system.

Now addressing the subscribers, the ILEC conducts loop qualification for each subscriber, insuring that the required service parameters in terms of bandwidth may be met. The ILEC also supplies the CPE, supplied by the ISP. Depending on the service deployment scenario, the subscriber may collect the CPE, the ILEC may ship the CPE, or as a last alternative, the ILEC will install on-site.

This last option, involving a *truck-roll*, is the least desirable in terms of impact upon the revenue model. Upon CPE installation, the ILEC will verify proper operation between the ADSL CPE and the DSLAM. With this phase complete, the ILEC may now hand over the subscriber to the ISP.

5.1.3 CPE Installation and Configuration

The ILEC will deploy CPE supporting either bridging or routing depending upon the subscriber and the DSLAM. In the case of bridged subscribers, the ILEC deploys the Alcatel

ATU-R, requiring very little configuration. For PPP subscribers, the ILEC deploys the Cisco 677 DMT2 ADSL router, also requiring little configuration. Configuration is as follows: (Note that all Layer 3 address parameters are dynamically assigned at the time of PPP session establishment.)

```
#set dhcp server enabled
      supplies IP addresses to attached PCs from local address pool
      (beginning with 10.0.0.1)

#set nat enabled
      translates between local address pool and
      ADSL interface IP address assigned via IPCP

#set ppp wan0-0 ipcp 0.0.0.0
      ip address assigned by upstream router acting as a dhcp server

#set ppp wan0-0 dns 0.0.0.0
      automatic negotiation of primary and secondary dns addresses - cached
```

The situation is a bit different with the 3Com OfficeConnect 811 serving small business subscribers. This device requires some Layer 3 configuration, and although controlled by the ILEC, Layer 3 parameters are supplied by the ISP.

Figure 5.3 depicts the global CPE configuration as to whether it is operating in RFC-1483 or PPP mode. Here, "newco," the small business subscriber, is using RFC-1483 routing.

FIGURE 5.3.

General Configuration

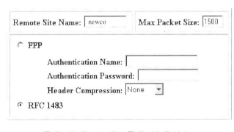

Figure 5.4 introduces the ATM configuration for the ADSL link. This includes the choice of PVCs or SVCs as well as their QoS. Here, newco is using a VBR-rt PVC. The CLEC providing the DSL service will likely configure these parameters, or it will instruct the ISP in question to perform the configuration.

FIGURE 5.4.

ATM Configuration

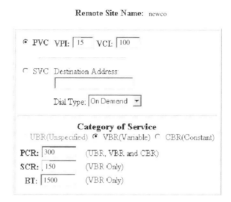

IP configuration is next (Figure 5.5), with the local and remote IP addresses of the link as well as the routing protocol in use (if any). The ISP has control over this phase of subscriber configuration.

FIGURE 5.5.

IP Configuration

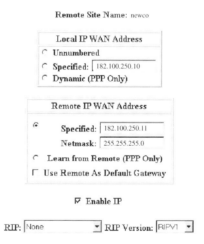

Finally, additional capabilities, unused here, include source validation and address translation (Figure 5.6). The use of NAT (Network Address Translation) or PAT (Port Address Translation) would be useful if the site used a private IP address scheme or wished to connect multiple hosts via the same public IP address.

FIGURE **5.6.**

Advanced Features

Remote Site Name: newco

IP Source Validation

Framed Routes
Manage

Address Translation
⊙ None
○ PAT
    Default Address: 0.0.0.0
        Manage Port Tables:
        Static TCP   Static UDP
○ NAT
    Manage Address Tables:
        Dynamic   Static

5.1.4 DSLAM Installation and Configuration

The ILEC will install and configure the DSLAM after conducting a Central Office site survey. The survey will take into account power, cooling, cabling, and future growth requirements. The incoming copper loops will first arrive at the Main Distribution Frame (MDF), where those belonging to ADSL subscribers are diverted to the POTS splitter in the ADSL rack. This separates the baseband POTS traffic from the ADSL data, forwarding the former to the narrowband voice switch and the latter to the ADSL modem shelf.

Note that different DSLAMs impose different VPI/VCI addressing requirements on the subscribers as well as on the ATM uplinks. In most cases, all subscribers will appear on VPI=1 and VCI=1 to n depending on whether a subscriber has been assigned more than one PVC.

At least one ATU-R, from Alcatel, assigns VCIs in round-robin fashion based on the ATM service category of the PVC. Within the DSLAM, these subscriber VPI/VCI pairs are remapped, with multiple subscribers sharing the same VPI (once again, dependent on the DSLAM in question).

If the provider has deployed subtending, there is an added level of remapping, and close attention should be paid to the vendor's guidelines. In addition, most implementations preclude or warn against the use of VCI=0–31 on the trunk side. This is due to the fact that some ATM switches do not permit VCIs in this range, and avoiding 0-31 will also help preclude the need for renumbering for future SVC deployment.

5

If the DSLAMs connect to an ATM access network, remapping of the VCIs and/or VPIs will occur within these switches as well. In all the examples, the VPI/VCI pairs depicted are associated with the last hop between the DSLAM or ATM switch and the service aggregator.

5.1.5 Service Aggregator Installation and Configuration

Now shifting attention to the ISP, Shastanets first deploys the service aggregator in a selected PoP. Although this is a less complex process than that carried out by the ILEC, Shastanets must still ensure that the proper physical requirements of the aggregator in terms of rack space, power (AC or DC), and cooling are met.

The ISP must also arrange with the ILEC for provisioning of the ATM trunk and coordinate for the communication of all PVC data. This process may be automated or manual. Note that basic CPE configuration is normally within the domain of the ILEC. The one exception here is the Cisco 1416, supplying routed services to the small business.

With this phase complete, Shastanets may now configure the Layer 2 (ATM) parameters of the aggregator for its subscribers as well as for its outgoing trunks. The ILEC communicates to Shastanets.net on which VPI/VCI pairs each subscriber will appear, as well as their ATM QoSs. Shastanets deploys two trunks from the Los Gatos aggregator for diversity: one to Santa Cruz and one to Sunnyvale.

```
Connection identifier:                      los_gatos-9-1-1
            ISP name:                       shastanets.net
            Node/slot/port:                 los_gatos/9/1
            VPI/VCI:                         30/80
            Interface encapsulation:         1483-R-LLC
            Layer 2 MTU:                     4096 (bytes)
            Connection type:                Trunk
            Layer 2 ATM service type:        GFR
            Traffic management (xmt/rcv):    PCR=100Mbps; MCR=45Mbps;
                                             MBS=4000 bytes;
```

Likewise, the parameters that follow are relevant to the access connections. Here, parameters for 1483 bridged, 1483 routed, and PPP connections are listed. Later, Shastanets.net associates these with subscribers.

```
Connection identifier:                      los_gatos_b_0001
            ISP name:                       shastanets.net
            Node/slot/port:                 los_gatos/9/3
            VPI/VCI:                         60/50
            Interface encapsulation:         1483-B-LLC
            Layer 2 MTU:                     1500
            Connection type:                Access
```

```
          Layer 2 ATM service type:       GFR
          Traffic management (xmt/rcv):   PCR=1Mbps; MCR=384Kbps; MBS=1500
                                            bytes
Connection identifier:                    los_gatos_ppp_0001
          ISP name:                       shastanets.net
          Node/slot/port:                 los_gatos/9/3
          VPI/VCI:                        7/200
          Interface encapsulation:        PPP/AAL5
          Layer 2 MTU:                    1500
          Connection type:                Access
          Layer 2 ATM service type:       GFR
          Traffic management (xmt):       PCR=1.2Mbps; MCR=512Kbps; MBS=1500
                                            bytes
                          (rcv):          PRC=300Kbps; MCR=150Kbps; MBS=1500
                                            bytes
Connection identifier:                    los_gatos_r_0001
          ISP name:                       shastanets.net
          Node/slot/port:                 los_gatos/9/3
          VPI/VCI:                        15/100
          Interface encapsulation:        1483-R-LLC
          Layer 2 MTU:                    1500
          Connection type:                Access
          Layer 2 ATM service type:       VBR-rt
          Traffic management (xmt/rcv):   PCR=300Kbps; SCR=150Kbps;
                                            MBS=1500 bytes
```

Since Shastanets is working with a single ADSL transport provider, all subscriber connections appear on slot 9, port 3. Nothing precludes the ISP from aggregating subscribers from multiple ADSL offerings. In this case, they would normally appear on different ATM access interfaces unless Shastanets itself operated an ATM access network for link aggregation. Figure 5.7 details these connections as configured.

FIGURE 5.7.

Los Gatos Connections

Next, Shastanets further configures the parameters of the two trunk interfaces, including encapsulations and QoS (including bandwidth) (Figure 5.8). Both of these trunks rely on RFC-1483 routed encapsulation.

FIGURE 5.8.

Los Gatos Trunks

Shastanets will also configure its RADIUS, PPP, and DHCP parameters. RADIUS configuration includes the server addresses/names, and whether they are used for authentication
and/or authorization, and the ports in use (Figures 5.9 and 5.10).

FIGURE 5.9.

RADIUS Default
Configuration

FIGURE 5.10.

RADIUS Server
Address

PPP configuration includes session timeouts and the authentication mechanism (PAP and
CHAP) (Figure 5.11), while DHCP includes server names/addresses.

FIGURE **5.11.**

PPP Default
Configuration

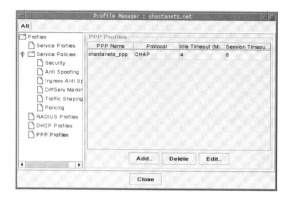

The ISP now configures the Layer 3 parameters for the trunks in use. These include the encapsulation, as well as the local and remote IP addresses.

```
Connection identifier                    los_gatos-9-1-1
     Interface                           gatos_santacruz
     Encapsulation type                  1483-R-LLC
     IP properties
          Local IP address and mask      182.100.30.1, 255.255.255.0
          Remote peers' IP address       182.100.30.2
Connection identifier                    los_gatos-9-2-1
     Interface                           gatos_sunnyvale
     Encapsulation type                  1483-R-LLC
     IP properties
          Local IP address and mask      182.100.32.1, 255.255.255.0
          Remote peers' IP address       182.100.32.2
```

Both trunks rely on RFC-1483 routed encapsulation, with one directed at Santa Cruz and the other to Sunnyvale.

More interesting is subscriber provisioning, where the ISP configures the Layer 3 parameters for each of the subscriber connections. Note the combination of encapsulation types: RFC-1483 bridged, routed, and PPP over ATM.

In this example, the service aggregator is used for simple DSL PVC termination and aggregation. Refer to Section 5.5 for more sophisticated firewalling and traffic management features. Subscriber parameters include whether authentication is required (yes, in the case of PPP), the local and remote IP addresses, and whether there is a LAN with separate IP addressing behind the CPE.

```
Connection identifier                    los_gatos_b_0001
     Subscriber                          burrito
     Encapsulation type                  1483-B-LLC
     Authentication                      no
```

```
        IP properties
            Local IP address and mask          182.100.210.101
            Remote peers' IP address and mask  182.100.210.102 255.255.255.0
Connection identifier                          los_gatos_b_0002
    Subscriber                                 friars
    Encapsulation type                         1483-B-LLC
    Authentication                             no
    IP properties
        Local IP address and mask              182.100.210.101
        Remote peers' IP address and mask      182.100.210.106 255.255.255.0
Connection identifier                          los_gatos_ppp_0001
    Subscriber                                 library
    Encapsulation type                         PPP/AAL5-LLC
    Authentication                             yes
        Username                               library
    IP properties
        Local IP address and mask              unnumbered
        Remote peers' IP address and mask      dynamic
Connection identifier                          los_gatos_ppp_0002
    Subscriber                                 dave
    Encapsulation type                         PPP/AAL5-LLC
    Authentication                             yes
        Username                               dave
    IP properties
        Local IP address and mask              unnumbered
        Remote peers' IP address and mask      dynamic
Connection identifier                          los_gatos_ppp_0003
    Subscriber                                 cath
    Encapsulation type                         PPP/AAL5-LLC
    Authentication                             yes
        Username                               cath
    IP properties
        Local IP address and mask              unnumbered
        Remote peers' IP address and mask      dynamic
    Reachability
Connection identifier                          los_gatos_r_0001
    Subscriber                                 newco
    Encapsulation type                         1483-R-LLC
    Authentication                             no
    IP properties
        Local IP address and mask              182.100.250.10
        Remote peers' IP address and mask      182.100.250.11 255.255.255.0
    Reachability                               215.110.37.0
```

The first two subscribers, "burrito" and "friars," are RFC-1483 bridged and thus share a single bridge group as well as the same IP address at the SSG side of the connections. Their addresses are all preassigned, and they require no authentication.

The PPP subscribers, "library," "dave," and "cath," all rely on dynamic addressing. Finally, the one RFC-1483 routed subscriber is a small business; "newco" relies on preassigned IP addresses and also has a LAN behind the ADSL router that must be reachable from the Internet.

The concept of a *bridge group* is important when working with RFC-1483 bridged subscribers. Here, a number of users may share a single Layer 2 domain; a topology, in effect an extension of the LAN environment.

Within a bridge group, leakage of unicast and multicast traffic is a factor of the vendor's bridging implementation. In this example, the bridged subscriber, "friars," shares the same bridge group as "burrito." Both appear at the SSG on 182.100.210.101. Addresses assigned to this bridge group span from 182.100.210.102 to 182.100.210.105. Figure 5.12 depicts subscriber connection configuration, including entry of bridge group parameters.

FIGURE 5.12.

Subscriber Configuration

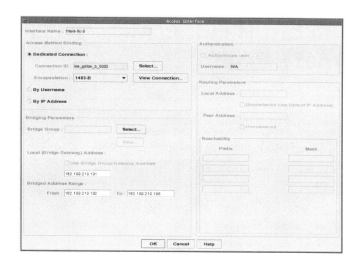

These subscriber configurations result in the table shown in Figure 5.13. Note that the services field is blank. Here, Shastanets.net provides only basic ADSL aggregation.

FIGURE **5.13.**

Shastanets' Los Gatos
Subscribers

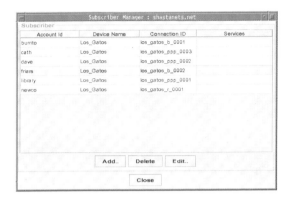

5.2 The Last Hundred Meters: PPP over Ethernet (PPPoE)

Scenario: A residential subscriber wishes to attach two PCs to the ADSL service, each connected to a separate upstream destination. For example, the husband or wife telecommutes, and requires a connection to the corporate gateway, while the kids access the Internet via a well-known ISP. Any solution must be simple enough to be implemented by both the DSL provider and the subscriber, must be secure, and must not require any infrastructure swapout.

Solution: In summer 1998, the ADSL Forum began development of PPPoE. This protocol meets the solution criteria outlined previously and has been deployed by at least one major ISP. PPPoE relies on support within the PC (a client shim) as well as within the aggregator, and is detailed in Chapter 4. As an aside, an alternative to PPPoE is to create a PPTP tunnel between the PC and the ATU-R. The user's PPP session is carried within this tunnel to the ATU-R, where the tunnel terminates and the PPP session is forwarded over the ATM/ADSL loop to the service aggregator. This architecture is supported as part of the Microsoft Windows 98 environment and by at least one ATU-R vendor: Alcatel's SpeedTouch Home.

The service aggregator strips off the PPPoE shim, authenticates each user within the subscriber's household IAW the L2TP tunneling model, and forwards the traffic over the appropriate upstream connections to either the corporation or the ISP.

Topology: As shown in Figure 5.14, the subscriber's two PCs connect to the ATU-R via a hub. Traffic flows across the local loop, through the DSLAM, and into the SSG-5000 service aggregator. From the aggregator, the PPP traffic flows across the ATM regional network to both the corporation and the ISP.

FIGURE **5.14.**

PPPoE Topology

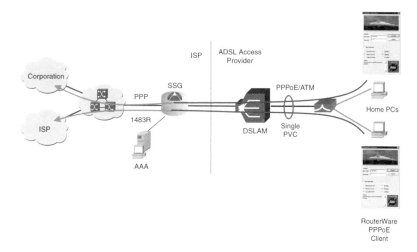

Configuration: The subscriber installs the RouterWare PPPoE client on each of the PCs. This client allows for the selection of multiple PPPoE sessions. Figure 5.15 depicts an installation where the subscriber may select between basic and premium service, as well as a connection to the corporate network.

Here, one user establishes a PPPoE session across the ADSL loop to the SSG, which then strips off the PPPoE shim and forwards the data as PPP to the corporation. The other user establishes a PPPoE session (sharing the same ATM PVC) to the SSG. Here, the PPPoE session terminates, the user is authenticated, and the traffic is routed to the destination ISP. The login procedure is identical to the one used within Windows dial-in. Figure 5.15 depicts the PPPoE client, showing selection of different QoSs/destinations.

5

FIGURE **5.15.**

*PPPoE Client
Application*

5.3 Beyond Aggregation

Scenario: An ISP wishes to offer value-added services to small businesses and some residential subscribers through an arrangement with the ADSL transport provider. In addition, it wishes to support multiple subscriber data encapsulations.

Solution: Building upon the simple aggregation example (discussed in Section 5.1), Shastanets.net applies a set of service profiles to the identified residential and business subscribers. These profiles are created by defining a set of policies such as firewalling, antispoofing, and DiffServ marking (where the ToS bits are set in the IP header).

5.3.1 DiffServ Marking

It is the profile that relates to a tariffed service in most cases, as opposed to the discrete policies. In this example, three of the subscribers, "newco," "dave," and "cath" are associated with profiles (Figure 5.16).

Topology: As in the first example, all subscribers connect to either ADSL routers or bridges, and then via DSLAMs to the service aggregator. This, in turn, connects to the provider's IP backbone.

FIGURE **5.16.**

Adding Value

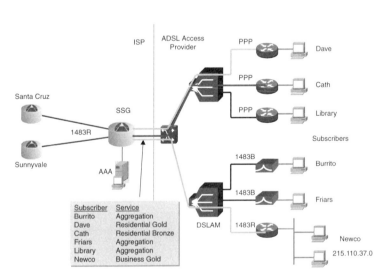

Next, Shastanets.net creates profiles and policy templates that are later applied to the subscribers. Profiles apply to all the ISP's subscribers, while policies are used to build tariffed services. Section 5.1 describes the creation of RADIUS, DHCP, and PPP profiles.

In contrast to ISP-wide profiles, policies relate to actions relating to the subscriber traffic. As listed in Figure 5.11, these include security (firewalling), antispoofing, and traffic management.

Configuration: Figure 5.17 depicts a traffic management policy for two different classes of subscribers relying on IETF DiffServ marking. Here, video and voice conferencing traffic (H.323) is painted with the highest Assured Forwarding class (AF4) and a Drop Priority of 0. All other traffic is painted with an AF3, the next highest priority.

FIGURE 5.17.

*Shastanets Gold
DiffServ Marking
Policy*

This is a traffic management policy suitable for a premium subscriber; thus the label is "gold," which will later be applied to service profiles belonging to the ISP's "gold" subscribers.

A step down is a DiffServ marking policy which applies only to real-time traffic. All other traffic is left unchanged (Figure 5.18). This is a policy appropriate to a standard, or "bronze" subscriber and will be revisited in section 5.3.4.

FIGURE 5.18.

*Shastanets Bronze
DiffServ Marking
Policy*

5

5.3.2 Antispoofing

Another important policy is *antispoofing*, which is preventing a subscriber outside of the subscriber's network from masquerading as an internal network user. Here, any traffic with an address internal to the subscriber's network but destined to the subscriber is dropped, while all other traffic is accepted (Figure 5.19).

FIGURE 5.19.

*Shastanets Antispoof-
ing Policy*

The inverse of antispoofing is *ingress antispoofing*, which prevents the individuals internal to a subscriber's network from masquerading as others. They would attempt this by sending traffic into the ISP's network with a source address from a subnet not belonging to the subscriber. Here, the policy accepts traffic into the network only if it is sourced with an IP address officially assigned to the subscriber's network. All other traffic is dropped (Figure 5.20). In most cases, the provider will enforce both ingress and egress antispoofing for a given subscriber.

FIGURE 5.20.

*Shastanets Ingress
Antispoofing Policy*

5.3.3 Firewalling

Possibly the most interesting type of policy is *firewalling*, which is allowing the provider to specify which traffic types are allowed and which are dropped. Here, as part of a small business policy (Figure 5.21), external traffic (from the network) destined to the subscriber's Web Server, Mail Server, or DNS Server is accepted.

Depending on the type of user and their business, different firewall policies are appropriate. For example, some businesses may disallow access to some types of streaming content or may block certain destinations. In the extreme, a policy may allow only very limited types of data (such as SMTP mail) to a small set of destinations. The service provider will of course work with the subscriber to develop the policy most appropriate.

The "!" relates to the concept of templating, where the actual server addresses are specified only upon subscriber provisioning. ICMP traffic is accepted, as well as any traffic internal to the subscriber and destined to the Internet. All other traffic is dropped.

FIGURE 5.21.

Shastanets Small Business Firewalling Policy

The residential firewalling policy (Figure 5.22) is simpler, dropping all traffic destined for the subscriber. This provides an increased level of security.

FIGURE 5.22.

Shastanets Residential Firewalling Policy

5.3.4 Combining Policies into Tariffed Services

5

Combinations of these policies are now packaged with service profiles, usually relating to tariffed services. Here, the DiffServ marking, antispoofing, ingress antispoofing, and small business firewalling are combined into a "business_gold" profile (Figure 5.23). This profile may then be applied against subscribers.

FIGURE 5.23.

Shastanets Business Gold Service Profile

Due to the differing requirements of residential subscribers, the firewall policy is altered for the "residential_gold" profile (Figure 5.24).

Finally, a "bronze" subscriber still uses the firewalling policy, but now it is marked in accordance with the "bronze" Diffserv profile described in section 5.3.1 (Figure 5.25).

Referencing the first example (Section 5.1), all trunk and subscriber configurations are identical in terms of Layer 2 parameters and Layer 3 addressing. The only difference here is that some of the subscribers are now assigned service profiles.

Figure 5.26 lists which service profiles have been assigned to which subscribers. The subscriber "dave" is associated with "residential_gold," "cath" with "residential_bronze," and "newco" with "business_gold." The subscriber "library" has been assigned a set of discrete unbundled policies—firewalling, antispoofing, and ingress antispoofing, but not Diffserv marking.

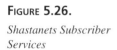

FIGURE 5.26.

*Shastanets Subscriber
Services*

Subscriber provisioning is now complete.

5.4 Internet Wholesaling

Scenario: A DSL access provider wishes to offer Internet wholesaling services to its sub-scribers, rather than forwarding all traffic at the ATM PVC level to upstream providers. The proposed service would allow subscribers to access multiple ISPs via the service aggrega-tor. Mappings between the subscribers and their primary ISPs are hardcoded within the ser-vice aggregator. The subscriber's traffic is then routed to its ISPs PoP where it is authenticated via RADIUS, at which time it has full access to its ISP. The solution should allow the ISP to manage its subscribers within the service aggregator, with the ADSL pro-vider responsible for the management of only the Layer 2 connectivity.

Solution: The ADSL provider deploys a service aggregator capable of running multiple vir-tual routers, one for each client ISP (Figure 5.27). The provider assigns groups of access and trunk VCCs to each ISP but has no visibility to the actual subscribers. The ISP then manages the Layer 3 parameters for each of its subscribers as well as its trunk connections. Each ISP client within the service aggregator has no visibility to any of the other ISPs, and the ADSL provider (the aggregator device owner) takes no part in IP address assignment or manage-ment.

5

Topology: Here, subscribers connect via ADSL routers and bridges to a DSLAM and then into the service aggregator. The aggregator then connects to the DSL provider's backbone which peers to multiple ISPs.

FIGURE 5.27.

Internet Wholesaling

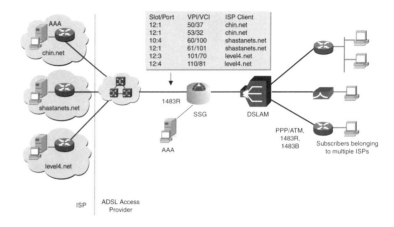

5.4.1 Subscriber Service Gateway Configuration

The SSG device owner (ADSL transport provider) will configure the set of ISPs that will be delivering services via the SSG. In this case, client ISPs include "chin.net," "level4.net," "shastanets.net," "skyline.net," and "various.net" (Figure 5.28).

FIGURE 5.28.

Internet Wholesaling ISP Clients

These ISPs operate as independent entities managing their own subscribers and routing tables via "virtual routing" entities, sometimes referred to as "contexts." This is critical in Internet wholesaling, where the transport provider eliminates the need for the individual

ISPs to terminate large numbers of ATM connections. Note the "device_owner" ISP: Nothing precludes the SSG device owner from delivering services to other ISPs while acting as its own ISP.

Looking more closely at the SSG, the device owner configures the Layer 2 parameters for each access (subscriber) and trunk connection. Figure 5.29 lists three of the client ISPs—chin.net, Shastanets.net, and level4.net—and their associated physical connection parameters. The Interface column states whether the Layer 3 parameters for the connection in question are configured. In this case, Shastanets.net has assigned subscribers to three access connections and has configured the two trunks.

FIGURE 5.29.

Sunnyvale Connections

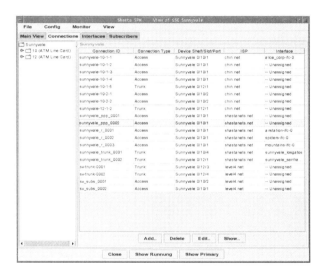

To complete the picture, a single ISP, in this case Shastanets.net, will deliver services via multiple service aggregators. Here, Shastanets.net is a client on the Sunnyvale, Los Gatos, Santa Cruz, and San Francisco SSGs (Figure 5.30).

FIGURE 5.30.

Shastanet.net's SSGs

5.4.2 ISP OSPF Configuration

Each ISP client is responsible for configuring its routing parameters, be they RIP, OSPF, or BGP. Here, a basic OSPF configuration is depicted.

```
Routing configuration
    General                            enable OSPF
    Static routes                      none
    Summary routes                     129.245.100.0/24
OSPF area configuration
    General
        Area ID:                       0
        Type:                          normal
        Cost:                          10
        Authentication type:           none
    Ranges:
        IP address:                    140.100.100.0/24
        Polling interval:              120 (default)
        DR eligible:                   yes
    Interfaces:
        General:
            DR priority:           1
            Cost:                  10
            Key:                            N/A (only if password-protected)
            Retransmission interval:     5 (default)
            Estimated transmission delay:  1 (default)
            Hello interval:             10 (default)
            Dead interval:              40 (default)
        Neighbors:
            IP address:                140.100.100.18/28
            DR eligible:               yes
            Polling interval:          120
    Virtual links:
        None in use
```

5.5 Portals

Portals are a subscriber's entry point into the network—their homepage for Internet access. They take two forms:

- The first class is the voluntary portals, such as www.netscape.com and www.yahoo.com, which sit and wait for subscriber access. The portal provider has no control over who visits the site and, in most cases, maintains no records.

- In contrast, mandatory portals are operated by specific service providers on behalf of their subscribers. They direct a subscriber to a designated homepage, listing services, destinations, and usually containing advertisements. They may also be a point at which the subscriber may configure characteristics of the service, such as firewall or VPN attributes.

FIGURE 5.31.

Mandatory Portal: Zoomtown

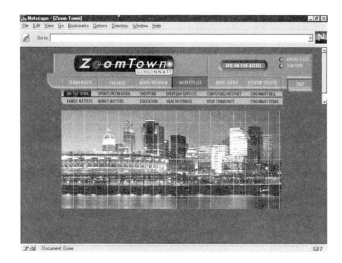

One of the first DSL-specific portals was deployed as part of Cincinnati Bell's Zoomtown Service (Figure 5.31), offering special content and services. Looking to Europe, Sonera in Finland offers an IP-based connectivity and multimedia service known as IP Communicator (Figure 5.32). Differing from other portal services, its focus is initially on voice communications.

5

FIGURE 5.32.

Voluntary Portal: Sonera (Finland)

FIGURE 5.33.

Mandatory Portal:
FreePC

A recent concept is the FreePC (Figure 5.33), where the desktop is customized with always-on advertisements The user, however, gets a free PC as part of the deal.

In all cases, subscribers of these services are automatically directed at the specific home-pages where vendors may advertise services or products. In addition, mandatory portal providers maintain subscriber databases: They know who accesses the network, when they access it, and what they do.

The portal concept actually applies to a number of the examples in this chapter. Bridged, PPP, PPPoE, tunneled, and VPN subscribers may all be directed to a service provider's portal. Here, the subscriber's HTTP traffic is initially intercepted by the SSG and directed to a portal server (1 in Figure 5.34).

FIGURE 5.34.

Portal Topology

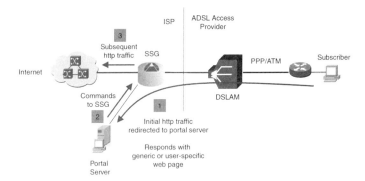

If the user has an entry within the provider's AAA database, the server may serve up a customized web page (2). Listed services may include destinations, subscription services, or content providers. If not, the subscriber will see a generic page. An example would be part of a free Internet service. The portal server now directs the SSG to send subsequent HTTP requests along the normal path (3). This procedure is also sometimes known as *service selection*. An additional element of service selection is whether a subscriber has access to a single destination or multiple destinations simultaneously. These issues were covered in Chapter 4.

5.6 Corporate Intranet Access: PPP/L2TP Tunneling

Scenario: Corporate telecommuters wish to access their corporate Intranet securely via the facilities of an ADSL provider. Ideally, they would be able to access the shared resources of an ADSL service provider, rather than requiring dedicated ATM VCCs back to the corporate gateway as in the previous example.

Solution: The ADSL provider implements a VPDN service, securely tunneling the subscriber PPP sessions from the provider's service aggregator acting as an L2TP Access Concentrator (LAC) to the subscriber's corporate L2TP Network Server (LNS). The subscriber's PPP session is placed in the proper L2TP tunnel via domain authentication with the ADSL provider's AAAA server. Chapter 4 describes this architecture in detail.

Topology: The subscriber is equipped with a telecommuter ADSL router, capable of PPP encapsulation. The ADSL provider operates the DSLAM, the LAC, and a MERIT RADIUS server. The corporation deploys a router capable of operating as an LNS. Here, the CPE router is a Cisco 675, the service aggregator is the Shasta SSG-5000, and the LNS is a Cisco 7206. CPE router configuration is identical (with the exception of IP addressing) to the first example (Figure 5.35).

FIGURE 5.35.

PPP/L2TP Topology

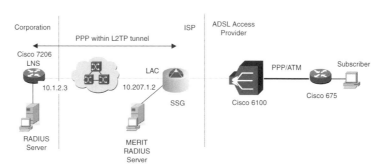

5.6.1 LAC Configuration

Subscriber PPP sessions arrive at the SSG-5000 via a trunk from the ADSL transport provider (the ILEC). During PPP session establishment, the subscriber inputs a fully qualified domain name. Based on this domain, as verified against a local AAA server, the PPP session is encapsulated within an L2TP tunnel. This tunnel terminates at an LNS located at the corporation. The complete procedure is described in greater detail within Chapter 4.

The following SSG configuration is formatted in the same style as the previous examples. Here, a given trunk PVC is associated with a specific L2TP tunnel. Note that multiple tunnels may of course follow the same physical path on different PVCs.

```
Connection identifier:                  trunk-0001
      ISP name:                         gatos.net
      Node/slot/port:                   adsl/4/1
      VPI/VCI:                          10/100
      Tunnel-ID                         1
      Interface encapsulation:          L2TP/AAL5
      Layer 2 MTU:                      4096
      Connection type:                  Trunk
      Layer 2 ATM service type:         GFR
      Traffic management parameters:    PCR=10Mbps; MCR=4Mbps; MBS=4000 bytes
```

5.6.2 Subscriber Configuration

Subscriber configuration is much like the previous examples:

```
Connection identifier:        sub-ppp-0001    ISP name:      gatos.net
      Confirmation key:       cats-access1
      Node/slot/port:         adsl/4/1
      VPI/VCI:                100/101
      Interface encapsulation: PPP/AAL5-LLC
      Layer 2 MTU:            1500
      Connection type:              Access
      Layer 2 ATM service type:     GFR
      Traffic management parameters:  PCR=1Mbps; MCR=384Kbps; MBS=1500 bytes
Connection identifier:        sub-ppp-0002    ISP name:      gatos.net
      Confirmation key:       cats-access2
      Node/slot/port:         adsl/4/2
      VPI/VCI:                101/101
      Interface encapsulation: PPP/AAL5-LLC
      Layer 2 MTU:            1500
      Connection type:        Access
      Layer 2 ATM service type:     GFR
      Traffic management parameters: PCR=1.2Mbps; MCR=512Kbps; MBS=1500 bytes
```

Layer 3 trunk configuration includes the tunnel destination, name, and security key.

```
Connection identifier                          trunk-0001
   Encapsulation type                          L2TP/AAL5
   IP properties
      Local IP address and mask                10.207.1.2/255.255.255.0
      Remote peers' IP addresses and masks     10.1.2.3/255.255.255.0
   L2TP properties
      Tunnel name (ASCII string)               foo-tunnel
      Password/Secret (is this as above?)      foo1
      Subscriber authentication                yes
```

Subscriber Layer 3 configuration is much like the previous examples, with encapsulation, security information, and IP addressing information. Note that the IP address of the subscriber is assigned by the AAA server located at the corporation:

```
Connection identifier                 sub-ppp-0001
   Encapsulation type                 PPP/AAL5-LLC
   Authentication                     yes
   Username                           ilikeppp
   IP properties
      Local IP address and mask       unnumbered
      Remote peers' IP address and mask   dynamic
Connection identifier                 sub-ppp-0002
   Encapsulation type                 PPP/AAL5-LLC
   Authentication                     yes
   Username                           ulikeppp
   IP properties
      Local IP address and mask       unnumbered
      Remote peers' IP address and mask   dynamic
```

The SSG administrator now creates an L2TP service profile, stating what groups of subscribers will utilize the VPDN service. During PPP session establishment, the RADIUS server returns a tunnel name for these subscribers. Their PPP session is then mapped into this tunnel.

5.6.3 LNS Configuration

The L2TP tunnels terminate at the LAC located at the corporate edge. With a router deployed as an LNS, the configuration will include the following tunnel-specific parameters:

- Definition of any locally defined users, including their domains and passwords. These are DSL subscribers forwarded across the L2TP tunnel. If an external AAA server is used, these local definitions are not required.

5

- Authorization information for the L2TP tunnel itself, including tunnel name, source and destination (LAC and LNS) hostnames, passwords, and encryption parameters (if available).
- Statement as to the type of tunnel if the platform supports multiple techniques (PPTP and L2F).
- ATM interface parameters, including the VCI/VPI in use for the tunnel, any traffic shaping, and IP addressing.
- A local IP address pool for tunnel users, if desired. If DHCP is deployed, this is not required.

The router configuration will of course include parameters regarding other interfaces in use, along with global routing definitions.

In most cases, the router vendor will provide on-line documentation for LNS (and LAC) configurations. The reader is urged to reference the relevant web pages on the different vendors' web sites for the most up-to-date configuration information, as commands, syntax, and product configurations change over time.

In this Cisco 7206 configuration file, RADIUS is used for subscriber authentication and accounting. When a new subscriber session arrives over the L2TP tunnel, the 7206 will check the RADIUS database to accept or reject the connection. If accepted, the subscriber is assigned an IP address from the telecommuters address pool.

The command "accept dial-in any virtual-template 1 remote dsl-server" allows the LNS to accept either L2TP or L2F tunneling. It associates tunnel users with virtual-template 1, which specifies the IP address pool and the type of authentication used.

It also specifies the name of the LAC—dsl-server. During tunnel establishment, the LNS sends its name (corporate-gateway) back to the LAC. The tunnel is carried over the single ATM PVC defined, while all traffic leaving the LNS is forwarded into the corporate backbone via the Fast Ethernet interface.

```
!
version 11.3
no service password-encryption
!
hostname corporate-gateway
!
aaa new-model
aaa authentication ppp default radius
radius-server host 10.18.30.1 auth-port 1645 acct-port 1646
radius-server key auth1
!
username corporate-gateway password 0 foo1
```

```
ip domain-name foo.com
ip name-server 171.69.2.132
ip name-server 198.92.30.32
ip multicast rpf-check-interval 0
vpdn enable
vpdn-group 1
 accept dialin any virtual-template 1 remote dsl-server
!
!
interface Ethernet0
 ip address 10.2.3.26 255.255.0.0
 no ip mroute-cache
 media-type 10BaseT
 no cdp enable
!
interface Virtual-Template1
 ip unnumbered Ethernet0
 no ip route-cache
 no ip mroute-cache
 peer default ip address pool telecommuters
 ppp authentication chap pap
!
interface ATM0
 no ip address
!
interface ATM0.123
 ip address 10.1.2.3 255.255.255.0
 pvc 0/144
  protocol ip 10.207.1.2 broadcast
  exit
 !
!
router igrp 99
 redistribute connected
 network 10.0.0.0
!
ip local pool telecommuters 10.36.1.1 10.36.1.255
no ip classless
ip route 0.0.0.0 0.0.0.0 Ethernet0
```

VPDN configuration is now complete.

5.7 Corporate Extranet Connectivity: VPRNs

Scenario: A second form of VPN, the Virtual Private Routed Network (VPRN), allows sites within a corporation to communicate securely across a shared infrastructure. Note that a type of VPRN initiated by the CPE has existed as part of managed router services for years.

5

In this configuration, the CPE routers, dedicated to a single customer, connect to a Frame Relay or ATM core. Using this architecture, any hierarchical routing requires traffic to exit the provider's cloud at a subscriber's site, or more commonly, to exit at a dedicated router for the customer provisioned by the managed router service provider. Both solutions are suboptimal in terms of routing and hardware costs.

Here, a network-based VPRN is provisioned by the service provider. The subscriber deploys simple CPE, with the service aggregator taking on the responsibility of maintaining separate forwarding tables persubscriber. Links across the provider's backbone are secured via IPSec.

In this example, a subscriber requires secure connectivity between DSL-connected branch offices and a small number of ATM-connected corporate sites. The requirement also exists for the branch offices to have direct Internet connectivity without having to traverse one of the central sites.

5.7.1 IP VPN Requirements

For effective deployment, IP VPNs introduce a number of requirements on the provider's physical and management infrastructure. These include (Gleeson, 1999):

- Local IP addressing
- Nonunique, overlapping addressing
- Data security: authenticity, privacy, and integrity
- QoS guarantees including bandwidth and latency
- Intra-VPN routing

Generic requirements include VPN identification, membership determination, reachability determination, advertisement of reachability, and the choice of a tunneling protocol.

Identification may be based on a globally unique VPN identifier. One proposal currently under debate within the IETF defines a 7 byte identifier consisting of a three byte IEEE OUI and a 4 byte index. An implementation of this for IP VPNs would be as follows:

```
[IETF OUI - 3 bytes][ISPs AS number - 2 bytes][VPN index - 2 bytes]
```

The next concern is VPN membership. What is the best way to identify subscribers and which VPN(s) an end-point belongs to. This may be accomplished via directory lookup (LDAP) or via network management. The same applies to other end-points in the same VPN. An additional option here is piggybacking into existing backbone routing protocols such as BGP-4.

Determining end-point reachability and the advertisement of this reachability may be via manual configuration, via an IGP such as OSPF, or via MPLS LDPs. A VPN end-point will learn which prefixes are reachable via the VPN through a VPN-specific IGP instance possibly through piggybacking in a backbone routing protocol.

5.7.2 VPRN Deployment

One complexity in VPRN deployment is whether branch offices have access to general Internet connectivity, and whether a single site may connect to multiple VPRNs. There are therefore four possibilities in order of increasing complexity:

- All sites connect to a single VPRN using public addressing.

  This scenario relies on the use of a single common routing table for all forwarding. This is much like conventional routing except that all traffic is encapsulated within IPSec tunnels. The downside of this solution is the requirement that all sites use global addressing, although nothing precludes the use of NAT at each site.

- All sites connect to a single VPRN, with Internet connectivity via a hub site.

  This scenario introduces per subscriber Forwarding Information Bases (FIBs). Traffic originating at a subscriber's site is forwarded across the backbone based on the routing entry within the FIB. This also permits use of overlapping or private address spaces, since each routing table operates independently.

  As with the first scenario, IPSec provides for security across the backbone. Traffic destined for the Internet is forwarded to one location, where it exits the VPRN and passes through a firewall. This is accomplished through the use of a default route, which points all sites to the location with the Internet connection.

- All sites connect to a single VPRN, and all (or most) sites have Internet connectivity.

  This scenario extends Internet connectivity to most or all sites. Here, traffic originating at a site is routed based on whether it is destined for a location within the VPRN. If a VPRN destination, this will take precedence.

- All sites connect to multiple, overlapping VPRNs, and may or may not have Internet connectivity.

  This scenario allows a given site to be a member of multiple VPRNs. An example here would be communities of interest within a large corporation, extranet connectivity, or traffic engineering. Selection of the proper per VPRN FIB can be based on the use of separate domains for each VPRN or, alternatively, based on destination addressing by looking up the FIBs in series.

5

Solution: The ISP deploys a VPRN service connecting the various subscriber locations. Shasta Networks SSG-5000s aggregate both ADSL as well as ATM traffic, forwarding it across the common backbone. DSL subscribers connect into DSLAMs and then the SSG, while ATM subscribers connect via an ATM switch. Within the SSG, a separate routing table is maintained on a per VPRN basis. Although many of the subscriber requirements could be met through the use of an L2TP-based VPDN, VPRNs are more suitable for environments where a branch must connect to more than one central site.

Topology: Figure 5.36 depicts the VPRN topology, with two subscribers, each sharing the same backbone. Traffic arrives at the SSG from a branch office or headquarters site. Here, based on the subscriber identity, the proper Forwarding Information Base (FIB) is used to determine the IP next hop.

FIGURE 5.36.

VPRN Topology

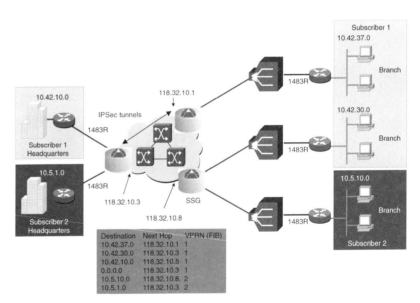

Note that subscriber 1 uses its headquarters site as the default route to the Internet. Also note the use of overlapping address spaces: This is permitted since the SSGs maintain a separate routing table (FIB) for each subscriber.

5.7.3 Service Aggregator Configuration

Basic subscriber configuration is much like the previous examples, with all sites belonging to a given subscriber relying on RFC-1483 routed encapsulation. Sites belonging to VPRN1 rely on the headquarters site for Internet connectivity, while VPRN2 permits Internet connectivity at each location. Within the SSG, site traffic is mapped to a specific VPRN based on preconfiguration.

The system maintains a separate Forwarding Information Base (FIB) for each VPRN. Here, the table lists the destinations and next hop IP addresses for each of the two configured VPRNs. Traffic between the SSGs is IPSec encapsulated. The current IPSec profile maintained by the ISP is depicted in Figure 5.37. Here, the Encapsulating Security Payload (ESP) is set to Triple-DES for encryption, and rekeying relies on Message Digest: 5. The Lifetime is the amount of time or the quantity of data for which the key is valid.

FIGURE 5.37.

IPSec Profile

The ISP creates a service profile whereby all traffic belonging to this subscriber is forwarded into the VPRN. The SSGs automatically handle intra-VPRN connectivity and security.

5.8 PSTN Bypass: VoIP

Scenario: A residential user wishes to use the ADSL connection for both data and VoIP traffic and is now relying on the ISP for voice transport.

Solution: The ADSL provider deploys ADSL modems integrating both Ethernet and a POTS interfaces, the latter converting the POTS traffic to VoIP. This modem implements the necessary QoS mechanisms to prioritize the VoIP traffic properly, transmitting it over a separate ATM VBR-rt VCC from the best-effort data traffic, carried over a GFR (UBR+) VCC.

Note that use of a separate VCC for the VoIP traffic is currently a requirement (as described in Chapter 4), since the DSLAM may only prioritize traffic at the ATM Layer. The DSLAM would have no way to prioritize traffic within a single ATM VCC.

The DSLAM then forwards the traffic to the SSG. Here, all ATM VCCs are terminated, and the SSG marks the high- and low-priority traffic in accordance with IETF DiffServ. The high-priority VoIP traffic arriving on the VBR-rt VCC is marked to AF4, while the remainder of the data traffic on the GFR (UBR+) VCC is marked to AF2.

Voice traffic from John to Anthony is forwarded over the SSG's Fast Ethernet interface. Its destination on the PSTN maps to the IP address of the VoIP gateway via an E.164 to IP lookup table. The FE connection supports the DiffServ marking. In contrast, traffic from Sue to Arthur remains on the Internet backbone and must be handled accordingly. Here, the trunk from the SSG to the Internet supports two VCCs, each supporting a different QoS and mapped to a different AF class (see Figure 5.38).

FIGURE **5.38.**

VoIP Topology

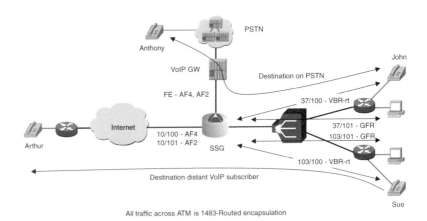

All traffic across ATM is 1483-Routed encapsulation

This example touches only on the signaling aspects of scalable VoIP transport from the standpoint of the CPE, the aggregator, the gateway, and the ISP backbone. As time progresses, it is expected that all devices, the service aggregator included, will implement increasingly sophisticated capabilities such as GSMP and may even integrate the protocol conversion, precluding the need for an external PSTN gateway.

Figure 5.38 depicts the VoIP topology, with two subscribers, John and Sue. Each has two PVCs, one for data (GFR) and one for the VoIP traffic (VBR-rt). At the SSG, the VoIP traffic is forwarded either across the backbone or to a local VoIP GW depending on whether the destination is local or distant. Here, John is calling a number reachable via the local gateway, while Sue's destination is remote.

The ATM trunk between the SSG and the Internet core supports two PVCs, each assigned a different AF, while the Fast Ethernet trunk from the SSG supports the two AFs via queuing.

Configuration: The ISP first configures the trunk and access connections. Here, the ISP defines three trunk VCCs: two for data traffic, and one for voice, as part of a single trunk identifier (lost_gatos-9-1-1). A second trunk connects to the Fast Ethernet port, directed at a local VoIP gateway. This trunk is used only for voice traffic.

```
Connection identifier:              los_gatos-9-1-1
    ISP name:                       shastanets.net
    Node/slot/port:                 los_gatos/9/1
    VPI/VCI:                        15/30
    Interface encapsulation:        1483-R-LLC
    Layer 2 MTU:                    4096
    Connection type:                Trunk
    Layer 2 ATM service type:       VBR-rt
    Traffic management parameter:   PCR=4000Kbps; SMR=2000Kbps;
```

```
MBS=128Kb
      Node/slot/port:                    los_gatos/9/1
      VPI/VCI:                           105/31
      Interface encapsulation:           1483-R-LLC
      Layer 2 MTU:                       4096
      Connection type:                   Trunk
      Layer 2 ATM service type:          GFR
      Traffic management parameter:      PCR=10000Kbps; MCR=2000Kbps;
MBS=128Kb
Connection identifier:                   los-gatos-13-1-1
      ISP name:                          shastanets.net
      Node/slot/port:                    los_gatos/13/1
      Layer 2 MTU:                       1500
      Connection type:                   Trunk
```

Figure 5.39 depicts the basic trunk configuration, showing the use of multiple PVC for the different AF classes.

FIGURE **5.39.**

*Shastanets Trunk
Diffserv Marking*

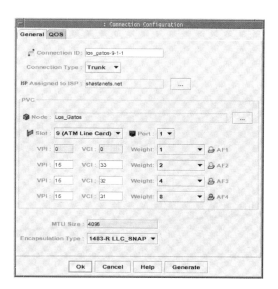

Figures 5.40 and 5.41 depict the QoS parameters for the voice and data VCCs across the first trunk (los_gatos-9-1-1). The voice traffic uses the VBR-rt VCC, while the data traffic rides over the GFR VCC.

FIGURE 5.40.

Shastanets Voice Trunk: VBR-rt

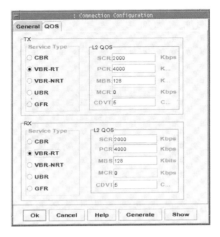

FIGURE 5.41.

Shastanets Data Trunk: GFR

Next, the ISP defines the subscriber (Figure 5.42), associating "newco" with a connection and encapsulation.

FIGURE **5.42.**

Shastanets VoIP
Subscriber

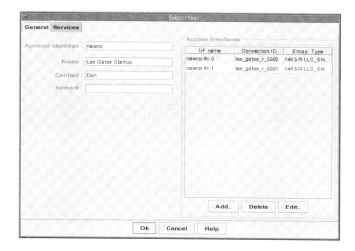

This draws on the two previously configured access PVCs for each of the QoSs, depicted in Figures 5.43 and 5.44.

FIGURE **5.43.**

Subscriber Voice PVC:
VBR-rt

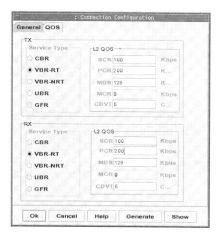

FIGURE **5.44.**

Subscriber Data PVC:
GFR

Layer 3 parameters for the trunks are listed as follows, including the local and peer IP addresses:

```
Connection identifier                        trunk-data1
   Encapsulation type                        1483-R-LLC
   IP properties
      Local IP address and mask              150.10.10.1
      Remote peers' IP addresses and masks   150.10.10.2
Connection identifier                        trunk-voice1
   Encapsulation type                        1483-R-LLC
   IP properties
      Local IP address and mask              150.10.11.1
      Remote peers' IP addresses and masks   150.10.11.2
Connection identifier                        trunk-voice2
   IP properties
      Local IP address and mask              150.10.101.1
      Remote peers' IP addresses and masks   150.10.101.2
```

Likewise, Layer 3 parameters for the subscriber connections include IP addressing, the encapsulation type, and whether authentication is required.

```
Connection identifier                        sub-0001-d
   Subscriber                                john
   Encapsulation type                        PPP/AAL5-LLC
   Authentication                            yes
      Username                               john
   IP properties
      Local IP address and mask              unnumbered
      Remote peers' IP address and mask      dynamic
Connection identifier                        sub-0001-v
   Subscriber                                john
   Encapsulation type                        1483-R-LLC
```

```
        Authentication                        no
        IP properties
           Local IP address and mask          unnumbered
           Remote peers' IP address and mask  129.245.200.11/24
Connection identifier                         sub-0002-d
     Subscriber                               john
     Encapsulation type                       PPP/AAL5-LLC
     Authentication                           yes
        Username                              sue
     IP properties
        Local IP address and mask             unnumbered
        Remote peers' IP address and mask     dynamic
Connection identifier                         sub-0001-v
     Subscriber                               sue
     Encapsulation type                       1483-R-LLC
     Authentication                           no
     IP properties
        Local IP address and mask             unnumbered
        Remote peers' IP address and mask     129.245.200.16/24
```

5.9 Entertainment: Video Streaming

Scenario: ADSL offers bandwidth sufficient to support video streaming. An ISP, realizing this revenue opportunity, collaborates with a content provider to deploy high-quality (MPEG-2) video streaming to a subset of the subscriber base. The application will, in effect, replace the corner video store. Note that this application is separate from existing web-based video streaming over IP at MPEG-1 quality and below.

Solution: The ISP colocates a Microsoft NetShow Theater video server with the service aggregator. Users establish on-demand ATM VCCs between the content server and their PCs or settop boxes for video traffic.

These SVCs follow Classical IP over ATM (RFC-1577) and require an ATMARP server for address resolution. Microsoft refers to this model as the Classical Internet Protocol (CLIP) in all documentation. Note that the use of SVCs implies that all networking devices along the path support or are transparent to ATM signaling. This includes the CPE router or bridge, the DSLAM, the service aggregator, and any ATM switches. The video content is MPEG-2 encoded and relies on UDP/IP over ATM AAL5 encapsulation. At the subscriber's PC, the Windows Media Player is used for playback. Depending on the hardware platform, a separate MPEG-2 decoder card may be required.

The control flow, between the subscriber's PC and the NetShow Theater Title Server, relies on TCP/IP. This title server communicates with the content servers and controls playback. A single video stream may be delivered by multiple content servers if the actual content is striped across disk drives on more than one system. This helps with load balancing and QoS, if multiple subscribers view the same content, and resiliency. The Microsoft Netshow web site lists supported client and server ATM NICs.

The ATM switch and service aggregator must support a feature known as funnel-join, in effect multipoint-to-point signaling allowing multiple content servers to merge their streams into a single subscriber SVC. Fore, Cabletron, and Cisco all have this capability as of the first half of 1999.

One important consideration is server scalability, a function of the number of discrete streams required, their bandwidth, and the total amount of content stored. As an example, a large system with 12 content servers in a fault-tolerant configuration will support 360 4 Mbps discrete streams, or the equivalent of 30 per server OC-3 interface. If the bandwidth was to be reduced to 2 Mbps per stream, the server could support 60 streams. At 4 Mbps, a single hour of content occupies 1.8 Gb of diskspace, doubled (3.6 Gb) if stored redundantly for fault tolerance. Five hundred films would therefore require on the order of 900 Gb of diskspace (nonredundant).

Unlike any of the previous examples, this solution relies on ATM SVCs. Therefore, the ATM VCC configuration on each of the devices is unique. The ATMARP server may reside on either a server or network device. For example, many ATM switches support this functionality.

Figure 5.45 depicts the video streaming topology, with a subscriber connected to both the Title Server as well as one or more Content Servers. Note that nothing precludes additional connections to the Internet.

FIGURE 5.45.

Video Streaming Topology

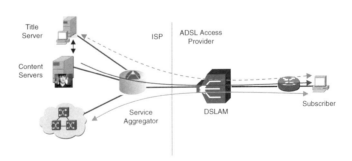

NetShow Theater configuration is as follows:

- Verify that all devices across the path support ATM (3.x/4.0) signaling or transparently pass signaling requests. Also verify that any ATM switches support funnel join and that all ATM NICs (client and server) are loaded with the proper NetShow drivers.

The service aggregator, an ATM switch in its own right, will support the required ATMARP function for the clients.

- Configure the MS NetShow Theater title and content server(s) based on scalability requirements.

- Connect the title and content server(s) to 155 Mbps ATM ports on the service aggregator. In this case, the servers are in physical proximity to the aggregator, allowing use of MM fiber. Given support on the aggregator, UTP5 would be an option as well.

- Subscriber PCs are equipped with integral ADSL modems supporting ATM SVCs based on RFC-1577 for both content and control flows. At the same time, the card supports PPP over ATM encapsulation for other (web, VoIP) traffic. Note that the client PCs must be configured with the well-known ATMARP server address.

5.10 Performance Testing

Scenario: This example is a bit different from the previous ones, in that it relates to testing the deployed hardware. With any ADSL deployment, there is a requirement to test the performance of the CPE, DSLAM, service aggregator, and any servers under load and under failure scenarios.

Here, focus is on the service aggregator and its capability to serve large numbers of ADSL subscribers with acceptable performance, as well as its ability to handle certain failure conditions gracefully. For example, a large percentage of the subscriber base may simultaneously attempt to re-establish the PPP sessions if the aggregator has restarted.

The aggregator and its attached RADIUS server must therefore be capable of servicing multiple authorization requests. Alternatively, during peak usage periods of the day (such as early evening), the aggregator must be capable of providing the necessary Layer 3 performance, whether routing or tunneling. Luckily, test tools capable of analyzing performance do exist and are used by vendors and service providers alike to verify proper operation of their deployments.

The goal here is to verify aggregator PPP performance as well as the capability of the aggregator and RADIUS server to service these subscribers properly in a timely manner.

Solution: The service aggregator, a Shasta Networks SSG-5000, will be attached to a Midnight Networks Avalanche PCON tool for performance testing. This test tool is used by a number of vendors for PPP over ATM performance analysis. Parameters tested may include

- Throughput
- Latency

5

- Total data traffic IP packets sent and received
- Total bytes sent and received
- Connection attempt successes versus failures
- Connection establishment or tear down speed
- Length of time connections remained up
- Time required and number of attempts to bring up or down LCP and IPCP
- LCP and IPCP negotiated options

The test system, a workstation with ATM and Ethernet interfaces, takes the place of the individual subscribers, and although only an approximation of real subscriber traffic patterns and behavior, deploying a test bed with thousands of PCs, is a bit unrealistic.

Use of PCON requires development of a test script, outlining how the test is to proceed. This includes the number of subscribers, level of PPP negotiation, security, and type of test traffic. For portability, the preferred method is to create two files: the first, an actual script describing the test, and a second, containing runtime variables for the first. The parameters of the test may therefore be altered on-demand.

Testing should be carried out with an awareness of RFC-1242 which describes benchmarking terminology and RFC-1944 which outlines methodology.

For example, the former document describes terms such as latency (which is the time interval for store and forward devices that starts when the last bit of the input frame reaches the input port and ends when the first bit of the output frame is seen on the output port) and back-to-back, (which is fixed length frames presented at a rate such that there is the minimum legal separation for a given medium between frames over a short to medium period of time, starting from an idle state). The latter document defines test topologies and includes typical frame sizes for testing.

Figure 5.46 details the PCON test topology. The PCON system connects to the SSG via an ATM interface. Traffic is forwarded to the SSG via this interface and returns to the PCON via a Fast Ethernet. A RADIUS server is also reachable via this interface.

FIGURE 5.46.

PCON Testing

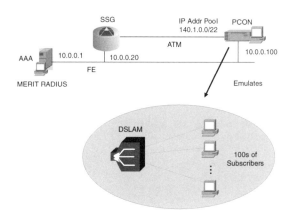

Configuration: Input to PCON consists of two files: The first is the actual test script; while the second includes the test parameters and is used by the first. Alternatively, all variables could be coded into the test script with a resulting loss in versatility.

The test script is as follows:

```
# THERE ARE NOT SUPPOSED TO BE ANY TWEAKABLE PARAMETERS IN THIS FILE!
# The variables you want to set are in the file 'datafile.tcl'.

proc pvctoport {n} {
   return [expr $n - 31]
}

proc porttopvc {n} {
   return [expr $n + 31]
}

proc dostatstable { fromport toport } {
   global async
   # print some stats
   puts "\n++pvc ¦ lcpstate ¦ ipcpstate ¦ lcptime ¦ ipcptime ¦
authstate ¦ port-pkts-sent ¦ port-pkts-recv ¦ ipl ¦ ipr"

 puts "++----------------------------------------------------------"
 for {set port $fromport} {$port <= $toport} {incr port} {
   set lcpstate [metric-get -port $async($port) -metric lcp-state]
   set ipcpstate [metric-get -port $async($port) -metric ipcp-state]
   set authstate [metric-get -port $async($port) -metric auth-state]
   set ipl [metric-get -port $async($port) -metric ip-address-local]
   set ipr [metric-get -port $async($port) -metric ip-address-remote]
   set lcptime [metric-get -port $async($port) -metric lcp-establish-time]
   set ipcptime [metric-get -port $async($port) -metric ipcp-establish-time]
   set portpktssent [metric-get -port $async($port) -metric port-pkts-sent]
   set portpktsrecv [metric-get -port $async($port) -metric port-pkts-recv]
     puts "++[porttopvc $port] ¦ $lcpstate ¦ $ipcpstate ¦ $lcptime ¦
```

```
$ipcptime ¦ $authstate ¦ $portpktssent ¦ $portpktsrecv ¦ $ipl ¦ $ipr"
   }
   puts "++"
   flush stdout
}

source randomprocs.tcl
if {[array exists async] != 1} {
puts "Adding avl stub function calls."
   source avlstubs.tcl
}

# global code starts here.

source single.data
random_init $seed 0
ra-ipaddress-set -port $eth(0) -ipaddress $etherip
port-macaddress-set -port 0-999 -macaddress $routermac

set firstport [pvctoport $firstpvc]
set lastport [expr $firstport + $simultaneous - 1]

ra-sleep $instantsleep

for {set repnumber 1} {$repnumber <= $totalreps} {incr repnumber} {

   if {$randomizeuseronpvc == 1} {
      set users [listshuf $users]
   }

   set pvclist {}
   for {set i 0} {$i < $simultaneous} {incr i} {
      lappend pvclist [expr $firstpvc + $i]
   }

   if {$randomizepvcs == 1} {
      set pvclist [listshuf $pvclist]
   }

   metrics-reset

   for {set outerloop 0} {$outerloop < $simultaneous} {} {
      for {set innerloop 0} \
         {($innerloop < $instantaneous) && ($outerloop < $simultaneous) } \
         {incr innerloop; incr outerloop} {

         set userno [expr $outerloop % [llength $users]]
         set user [lindex $users $userno]
         set port [pvctoport [lindex $pvclist $outerloop]]
```

```
                set authtype [lindex $user 2]
                if {$authtype == "rand"} {
                    set authtype [randomelement "pap chap"]
                }
                switch -- $authtype \
                    pap {
                    event-schedule ipcp-open -port $async($port) \
                        -timeout $ipcptimeout \
                        -pap-username [lindex $user 0] \
                        -pap-password [lindex $user 1]
                    } chap {
                    event-schedule ipcp-open -port $async($port) \
                        -timeout $ipcptimeout \
                        -chap-name [lindex $user 0] \
                        -chap-secret [lindex $user 1]
                    } both {
                    event-schedule ipcp-open -port $async($port) \
                        -timeout $ipcptimeout \
                        -pap-username [lindex $user 0] \
                        -pap-password [lindex $user 1] \
                        -chap-name [lindex $user 0] \
                        -chap-secret [lindex $user 1]
                    } default {
                        puts "$user has an unknown authtype."
                }
                puts "Scheduled pvc[porttopvc $port] $user using $authtype"
            }
        events-run
        ra-sleep $instantsleep
    }

    dostatstable $firstport $lastport

    for {set outerloop 0} {$outerloop < $simultaneous} {incr outerloop} {
        set port [pvctoport [lindex $pvclist $outerloop]]
        set ipcpstate [metric-get -port $async($port) -metric ipcp-state]
        if { $ipcpstate == 1 } {
            if { $trafficetoa == 1 } {
                set channel [ip-traffic-define -srcport $eth(0) -dstport
$async($port) \
                    -type $traffictype \
                    -count $trafficcount \
                    -size $trafficsize \
                    -rate $trafficrate ]
                event-schedule ip-traffic-recv -id $channel -timeout $traffi-
crecvtimeout
                event-schedule ip-traffic-send -id $channel
            }
            if { $trafficatoe == 1 } {
                set channel [ip-traffic-define -srcport $async($port) -dstport
```

```
$eth(0) \
                -type $traffictype \
                -count $trafficcount \
                -size $trafficsize \
                -rate $trafficrate ]
            event-schedule ip-traffic-recv -id $channel -timeout $traffi-
crecvtimeout
            event-schedule ip-traffic-send -id $channel
        }
    } else {
        puts "Skipping port $port, ipcp not up."
    }
  }
  events-run

  puts "Leaving them up for $uptime seconds"
  ra-sleep $uptime
  puts "Tearing them down."

  ra-sleep $teardownsleep
  ipcp-close -port $firstport-$lastport -timeout 30
  ra-sleep $teardownsleep
  lcp-close -port $firstport-$lastport -timeout 30
  ra-sleep $teardownsleep
}

puts "Done."
```

The previous script generates two tables: statistics and traffic. Figure 5.47 depicts the statistics output.

FIGURE 5.47.

PCON Statistics

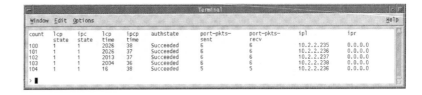

Descriptions of each of the displayed variables are shown in Table 5.1. Note that there are many additional IP, LCP, and IPCP level statistics gathered by the system, although not displayed in this example.

TABLE 5.1.   PCON Statistics Variables

Variable	Description
$lcpstate	Current state of LCP connection (0 = down; 1 = up)
$ipcpstate	Current state of the IPCP connection (0 = down; 1 = up)
$lcptime	Time required to establish the most recent LCP connection on the port (ms)
$ipcptime	Time required to establish the most recent IPCP connection on the port (ms)
$authstate	Current state of PAP or CHAP authentication (illegal, inactive, in-progress, succeeded)
$portpktssent	Number of packets sent on the port, including PPP, IP, IPCP, and LCP
$portpktsrecv	Number of packets received on the port, including PPP, IP, IPCP, and LCP
$ipl	Negotiated IP address on the PCON side of the connection
$ipr	Negotiated IP address on the DUT side of the connection
$ippktsrecv	Number of IP packets received on the most recent connection on the port
$ipbytesrecv	Number of IP bytes received on the most recent connection on the port
$ippktssent	Number of IP packets sent on the most recent connection on the port
$ipbytessent	Number of IP bytes sent on the most recent connection on the port

As noted previously, the script relies on a number of runtime variables contained in a separate file. A sample set of values and their explanations follows:

```
# Number of repetitions of full test to perform
set totalreps 1
# Starting pvc for connection range
set firstpvc 100
# Number of sessions to bring up before teardown
set simultaneous 1
# Number of sessions to bring up at a time
set instantaneous 4
# Number of ticks to wait between test groups
set instantsleep 1
# Leave the connections up for uptime seconds before teardown
set uptime 10
# Number of ticks to wait before and after each close command
set teardownsleep 0
# Timeout for ipcp-open command:
set ipcptimeout 15
# Seed for random number generator. [pid] == use process id.
#set seed [pid]
set seed 31337
# If set, bring up pvc's in random order
set randomizepvcs 0
```

5

```
# If set, pick random user for next pvc
set randomizeuseronpvc 0
# Set traffic type to pcon specific type:
set traffictype http
# Number of packets to send (one way)
set trafficcount 10
# Size of each packet sent in bytes (<~1460)
set trafficsize 1000
# transmission rate of each packet in pkts/second
set trafficrate 1
# total listen time for packet recieve command
set trafficrecvtimeout 15
# send traffic from ethernet to ATM
set trafficetoa 1
# send traffic from ATM to ethernet
set trafficatoe 1
# ethernet ip address
set etherip 150.1.1.2
# ethernet default router's mac address
set routermac 00:90:6f:f1:38:2b
# Set of all known user/password/authtype tuples.
#     {{username@domain} {password} {pap|chap|rand|both}}
set users {
    {{use0@lab.shastanets.com} {shasta} {both}}
    {{user0000@lab.shastanets.com} {shasta} {both}}
    {{billyray@lab.shastanets.com} {shasta} {both}}
    {{user1@a1234567890123456} {shasta} {both}}
    {{user11@a1234567890123456} {shasta} {both}}
    {{sub11@bryan-isp} {shasta} {both}}
    {{sub12@bryan-isp} {shasta} {both}}
    {{sub13@bryan-isp} {shasta} {both}}
}
```

Summary

The examples in this chapter have brought together into a single place concepts introduced earlier in this book. The various hardware standards and platforms, software protocols, and service architectures have provided the basis for the different residential, telecommuter, and business implementation examples.

These examples hopefully provide the reader with a starting point for actual deployment and thus conclude the core material of this book. The final chapter provides a quick over-view of other currently deployed access technologies, including the different variations of DSL.

Endnotes

Gleeeson, B., Lin, A., Heinanen, J., Armitage, G., and Malis, A., "A Framework for IP Based Virtual Private Networks" Internet Draft: <draft-gleeson-vpn-framework-01.txt>, March 1999.

5

Chapter 6

Alternatives to ADSL

This chapter details some of the alternatives to ADSL—the other DSLs, cable modems, and traditional POTS/ISDN dial-up options. Although ADSL is the emphasis of this book, a service provider may deploy multiple access technologies in order to complete a service portfolio. This is especially true with DSL due to the distance versus bandwidth limitations of the various encoding techniques.

Previous to any of the "modern" technologies based on advanced coding, older methods of delivering megabit connectivity over copper did exist. Traditional T1 (1.544 Mbps) links rely on Alternate Mark Inversion (AMI), whereas E1 at 2.048 Mbps uses the more sophisticated High Density Bipolar 3 (HDB3).

As you might guess, AMI encodes a single bit of data per analog signal. Because of this, its frequency extends to 1544 kHz, and it is limited to 3000–6000 feet between repeaters. Due to the costs of circuit engineering (such as removing bridged taps), hardware, and maintenance, the service providers could charge up to $2000 per month for this service. AMI and HDB3 are both four-wire technologies, so they split the upstream and downstream directions on different pairs.

The first higher-density encoding technique deployed was the 2 Binary, 1 Quaternary (2B1Q) encoding used on most ISDN BRIs. This encoding was required to achieve the 18K feet distance over 26 AWG wire desired by ISDN's developers. The existing AMI encoding, topping out at 160 kHz, would have resulted in too great a signal attenuation. 2B1Q sends two bits of data per analog signal and occupies 80 kHz of spectrum. This same encoding is used by IDSL, SDSL, and some older HDSL implementations. More recently, CAP and variants of DMT have found their way into HDSL.

As previously noted, the DSLs form a continuum, and the typical service provider may deploy more than one of these technologies as part of a service portfolio. Whereas ADSL services more residential subscribers than businesses, SDSL and HDSL are more suited to businesses. Where providers have deployed IDSL, it is primarily a telecommuter service. Just as important are the distance versus bandwidth limitations of the different technologies: SDSL and IDSL are capable of servicing users beyond the range of ADSL.

Figure 6.1 depicts the nominal ranges for the different technologies, including the Very High Speed Digital Subscriber Line (VDSL), which is not yet deployed on any large scale. Each of these technologies is described in more detail in the following sections.

FIGURE **6.1.**

Nominal Ranges for
DSL Technologies

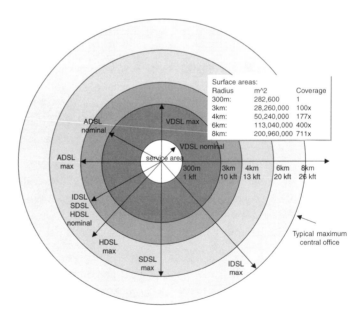

6.1 Integrated Digital Subscriber Line (IDSL)

The Integrated Digital Subscriber Line (IDSL) is a relatively simple technology that reuses the 2B1Q encoding used in ISDN, but without dividing the bandwidth into the 2B+1D data and signaling channels. Thus the entire 144 Kbps is available for symmetric data transport at distances of up to 26,000 feet. This is equivalent to the reach of ISDN, expected due to IDSL's reuse of ISDN encoding.

Unlike the other DSLs, which rely on standalone chassis-based solutions, the majority of early IDSL deployments relied on modules installed in existing D4 channel banks with the addition of framed DS1 uplinks. More recently, DSLAMs supporting IDSL, and frequently SDSL as well, have appeared.

IDSL has been most successful in North America, due to the installed base of D4 banks as well as the greater penetration of ISDN for data within Europe. Of note is a variant in Germany that permanently connects one ISDN B channel, much like a 64 Kbps IDSL service, while maintaining the second B channel and the D channel for dynamic connectivity. Still, deployment outside of the CLECs and PCLECs is not all that widespread, with only one of the RBOCs deploying the technology. Looking ahead, in the face of ADSL for residential services and SDSL for businesses, the future business case for IDSL is unknown. One possible use for IDSL is with DLC-attached subscribers where ADSL is not yet viable, since DLCs do not affect IDSL's encoding.

6.2 Symmetric Digital Subscriber Line (SDSL)

6

The Symmetric Digital Subscriber Line (SDSL) is in fact a classification that applies to a number of proprietary technologies enabling symmetric connectivity over a single copper pair.

Although the majority of SDSL implementations reuse ISDN's and IDSL's 2B1Q encoding, there has been no standardization effort along the likes of ADSL for higher level interoperability. Therefore, SDSL hardware from one vendor may not interoperate with that of another unless explicitly stated.

The technology commonly operates at data rates of 128, 192, 384, 512, 768, and 1152 Kbps and at distances of up to 12,000 feet. Additional distance is possible at the sacrifice of bandwidth, fitting well with the CLEC's business model of pushing service to the greatest number of users possible.

A variant of SDSL requiring two pairs pushes bandwidth to 1.544 Mbps. Thus maximum rate achievable via SDSL overlaps that of HDSL. In fact, with the deployment of HDSL2 requiring only a single pair for this bandwidth (described in a following section), the two technologies begin to blur.

As with ADSL, SDSL supports PPP, over both ATM and frame link layers depending on the vendor. Therefore, the tunneling and service selection services available within ADSL apply to this technology as well. In addition, ATM-based SDSL supports the service categories required for differentiated ATM services, given support within the SDSL DSLAM.

6.3 High-Speed Digital Subscriber Line (HDSL)

The High-Speed Digital Subscriber Line (HDSL) is currently the preferred method for delivering full T1/E1 connectivity to small businesses, replacing the older, more repeater-intensive (every 3 Kft) architectures of the past. Applications include leased lines, router interconnections, and PABX connectivity.

The first HDSL installations relied on 2B1Q encoding, pushing 784 Kbps over two wires or 1.5 Mbps (a full T1) over four wires to a maximum of 12 Kft (24 AWG). This maximum distance, not all that great, is a function of the 392 kHz upper bound of the frequency spectrum capable of being supported by the copper pairs.

E1 is even more constrained due to its higher bandwidth requirement and therefore originally required three copper pairs for its 2.048 Mbps signal. In newer HDSL installations, this requirement has been reduced to two pairs by relying on more advanced encoding such as CAP, which reduces the spectrum required and allows for reach of up to 18 Kft.

Figure 6.2 depicts HDSL CAP encoding in contrast to 2B1Q HDSL and the AMI encoding described earlier. The CAP encoding results in a greatly reduced frequency spectrum and therefore a greater reach.

FIGURE 6.2.

HDSL Frequency Spectrum

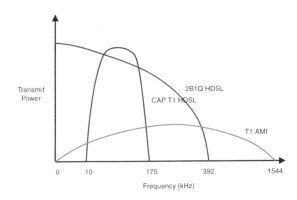

HDSL2 is an evolving standard, further refining the technology. Undergoing standardization within ANSI T1E1.4, HDSL2 is driven by a group of DSL vendors known as *Overlapped PAM Transmission with Interlocking Spectra* (OPTIS). The proposal is based on 8 PAM (Pulse Amplitude Modulation) and trellis modulation. In addition, it includes spectral shaping to help with co-existence in cable groups.

Benefits of HDSL2 include support for T1/E1 rates over single pairs, as well as an extension of the reach when deployed with two pairs. For example, whereas HDSL's 12 Kft limitation limits it to about 60 percent of the U.S. local loop and about 85 percent of the German local loop, HDSL2 extends this to over 15 Kft, pushing penetration to 78 percent in the United States and 92 percent in Germany (see Figure 6.3). Unlike HDSL, HDSL2 supports multiple transmission rates and has an option for POTS support.

Figure 6.3.

HDSL2 Pushes Penetration Up

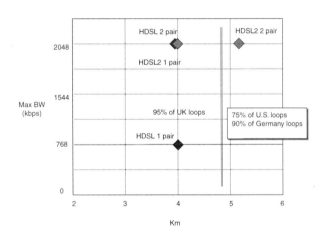

Another advanced HDSL solution is the wavelet Digital Subscriber Line (wDSL), proposed by Aware. It is capable of 1 Mbps over a single pair at distances up to 18 Kft, or 2 Mbps at 12 Kft. A variant relying on two pairs allows for 3–4 Mbps of throughput, while an added capability is POTS support in the baseband. This does not exist within standard HDSL, but it is a major selling point of ADSL. Although one could argue that the majority of HDSL subscribers, the small businesses, are not short on copper pairs, this is not always the case.

WDSL relies on a multicarrier technology in the same way as DMT, but extends this by implementing a new digital wavelet transform as opposed to DMT's use of the older Fourier transforms. This Discrete Wavelet Multitone (DWMT) results in more of the subchannel power being contained in the mainlobe at a given frequency. Figure 6.4 contrasts the two technologies, clearly showing the "crispness" of the DWMT power curve.

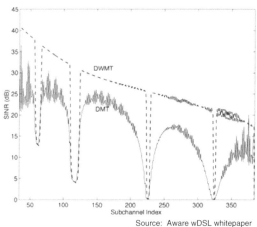

Source: Aware wDSL whitepaper

For example, whereas conventional DMT results in 91 percent of the transmitted power being in the mainlobe (with sidelobes 13 dB below the mainlobe), DWMT raises this to over 99.9 percent (with sidelobes 45 dB below the main). The net result is that the guard times implemented in DMT, which result in lower throughput, are not required within DWMT. Thus the technology has a greater capacity. There is no real reason why this encoding could not be implemented for ADSL as well.

6.4 Very High Speed Digital Subscriber Line (VDSL)

The Very High Speed Digital Subscriber Line (VDSL) is possibly the most interesting of the DSL alternatives. This is due to its potential to deliver true broadband applications to the subscriber by virtue of its downstream and upstream maximum data rates of 52 and 34 Mbps, respectively.

However, capability comes at a price in terms of distance, with VDSL deployed between a neighborhood or building fiber termination point and the end user. The point at which the fiber terminates within a Full Service Access Network (FSAN), first introduced in Chapter 3, goes by multiple names. If this termination point appears at the building, it is known as Fiber-to-the-Home (FTTH) or Fiber-to-the-Basement (FTTB). The fiber may also terminate at a street cabinet, known as Fiber-to-the-Node (FTTN) or at the curb, known as Fiber-to-the-Curb (FTTC). Alternatively, the fiber may terminate at the Central Office if subscribers are nearby (ADSL, 1998).

Although the first envisioned uses of VDSL were in an asymmetric configuration, symmetric VDSL exists as well. The first will find use for VoD, HDTV, and Internet access, whereas Inverse muxing and T1/E1 replacements between buildings and sites are expected to be major uses of the second. Here, VDSL will handle any applications currently served by these two technologies. Figures 6.5 and 6.6 depict proposed symmetric and asymmetric VDSL rates, respectively.

FIGURE 6.5.

Proposed Symmetric VDSL Rate

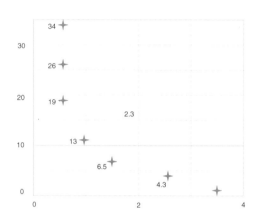

FIGURE 6.6.

Proposed Asymmetric VDSL Rate

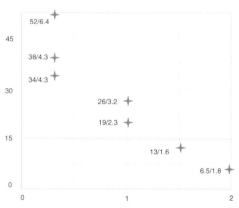

6

One of the major concerns with VDSL is interference with other signals in the same frequency spectrum and vice-versa. For example, AM broadcast (500 kHz to 1600 kHz), amateur radio bands (2 MHz, 3.5 MHz, 7 MHz, 10 MHz), and the HomePNA all overlap VDSL. The VDSL receiver must reject signals in these frequency ranges, and the transmitter must limit its power in these bands. These interferers are in addition to the crosstalk and attenuation which affect all the DSLs, but affect VDSL to a greater extent. The effect of VDSL on the other DSLs within a binder group is still an area for research.

In the United States, VDSL standardization is within the ANSI T1E1.4 group and began in 1995. Deliverables include a reference configuration, data rates, interworking, transmission and impairments, and OAM&P. Within Europe, ETSI's TM6 has released the first functional requirements and line code documents.

The ADSL Forum also provides some coordination in the form of a study group handling higher-layer protocols, customer interfaces, CPE and wiring, OAM&P, and network interfaces. Where VDSL is deployed for Internet connectivity, one may look at VDSL as part of a continuum of modem technologies with the same higher-layer protocol architectures— thus the interest in the ADSL Forum. The ADSL Forum is also investigating ADSL to VDSL migration scenarios in terms of equipment reuse, spectral compatibility, management, business drivers, and deployment within different fiber deployments (FTTH, FTTC, FTTN, and so on).

Finally, a multivendor alliance at www.vdsl.org helps to promote the technology. This group is separate from the VDSL group within FSAN, with a responsibility to model VDSL deployments and identify where standards are required (deferring to ANSI and ETSI for completion).

VDSL line codes are still somewhat in flux, with multiple proposals under consideration. Note that this is much like the earlier days of ADSL, where DMT, CAP, and QAM were all contenders. A group known as the VDSL Coalition, consisting of a number of the major semiconductor and telecommunications hardware providers, is proposing CAP/QAM (no different than that described in Chapter 2).

An alternative proposed by the VDSL Alliance is known as SDMT/Zipper. This alliance also consists of hardware vendors, while the encoding is a variation of DMT offering spectral compatibility with existing POTS/ISDN, HDSL, and ADSL services. Unlike ADSL's DMT, subcarriers alternate between upstream and downstream traffic. That is, if subcarrier 6 is upstream, 7 is downstream, and 8 is upstream. Looking a bit further afield, some vendor-specific solutions have been proposed to solve the distance vs bandwidth limitations of some DSLs.

6.5 Other DSLs

This section discusses other DSL options, including Multiple Virtual Lines (MVL) and the 1 Meg Modem from Nortel Networks.ok

A vendor-specific solution from Paradyne, Multiple Virtual Lines (MVL), is described here although it may be considered an ADSL solution due to its asymmetry. MVL is capable of data rates up to 768 Kbps and pushes the distance envelope to 24 Kft (7 km). As with other

DSLs, bandwidth is a function of distance. The technology operates in the 80 khz frequency spectrum as ISDN, and in fact supports voice without the use of splitters. Although a proprietary solution, MVL has found acceptance by a number of service providers seeking to offer DSL services to the largest subscriber base possible (refer back to the surface area calculations shown in Figure 6.1).

Yet another vendor solution is the 1 Meg Modem from Nortel Networks. This technology supports bandwidths up to 1.28 Mbps downstream and 320 Kbps upstream and is splitterless. One feature of the technology is that the modems are designed for installation in a carrier's existing DMS Line Concentrating Module (LCM) hardware, as opposed to relying on a separate chassis. Here, the 1 Meg modems replace the existing POTS line cards. This reliance on existing Nortel hardware may seem limiting, but within North America, the DMS holds a 50 percent marketshare. Unlike conventional DSLAMs with ATM or frame uplinks, the data uplink from the LCM is an Ethernet. This solution, of course, requires a custom CPE. Both standalone modems and PC NICs are available.

6.6 Cable Modems

No discussion of ADSL alternatives would be complete without mention of what is the largest and most publicized contender: cable modems. This technology has the backing of almost every global cable provider, and at the end of 1998 had outpaced ADSL (and all DSL) deployment by at least a factor of two.

Existing analyst reports such as Yankee, 1998, which predict 4.3 million subscribers in 2003 are predicted to be on the low side given interest by the cable providers. The concept behind cable modems is quite simple, although implementation is somewhat more complex. A coaxial cable entering a subscriber's home has a capacity for up to 1 GHz of spectrum, spread into broadcast channels whose band may be 6 MHz or also 8 MHz in some European regions.

Although all of these may be designated to video, the spectrum may be used for just about anything, including data. (Looking back in time, the old Wang broadband net, deployed during the mid-1980s, offered the same mix of video and data services on a more limited scale.)

To promote cable modem interoperability, a number of cable operators formed the Multimedia Cable Network System (MCNS) through CableLabs. The resulting standard is the Data-over-Cable Service Interface Specification (DOCSIS), incorporated within ITU-T J.112 Annex B, containing provisions for IP telephony, streaming video, VPNs, and QoS as well as simple data transfer.

Vendor CPE and headend products will certify to DOCSIS, guaranteeing interoperability. The DOCSIS specifies a physical and MAC protocol, offering asymmetric data service between the headend and the subscriber. Upstream data is managed by time slotting the upstream frequency into "minislots." The Cable Modem Termination Shelf (CMTS) scheduler informs the CPE modems of the minislot structure—which are contention-based, and granted to specific modems for real-time traffic. This relates to the QoS scheduling available within the DOCSIS 1.1 specification, where subscribers request QoS by describing a set of token bucket-like flow specification parameters. The system will then apply admission control to determine whether to accept the flow. If accepted, the CMTS assigns the cable modem a service identifier (SID), which describes which packets may be sent with this flow specification. Whether this complexity is actually required for VoIP services is yet to be determined.

In current installations, data occupies one or more of the channels, each of which supports on the order of 27 Mbps of traffic via QAM64 encoding. If the system is two-way enabled, the spectrum may be divided between downstream and upstream data. Here, the downstream spectrum is between 42 and 750 MHz, with the upstream data transmitted between 5 and 40 MHz. Since this section of the spectrum is noisy due to radio interference and cabling problems, Quadrature Phase-Shift Keying (QPSK) is the preferred upstream encoding, sacrificing bandwidth for additional resilience; 16QAM is also defined in the standard and can be used for upstream operation if the cable plant conditions permit. Under 16QAM, upstream throughput is on the order of 2–3 Mbps, while QPSK yields 1.5 Mbps. In this envrionment, the cable headend will usually designate from 6 to 8 upstream channels per single downstream channel for a set number of subscribers. This allows for almost equal upstream and downstream bandwidth. If the system is not two-way enabled, upstream data is transported by what is known as *telco-return*, a modem.

Figure 6.7 depicts a cable modem deployment, beginning with the headend, where the cable router connects to one or more coaxial links. It also connects to an ATM or packet backbone network, may include PSTN interworking (especially if the CPE routers support VoIP), and may interface to local IP video servers and web caches.

Note that the actual analog or digital broadcast signal does not pass through the router—instead it directly enters the distribution network. In the direction of the subscribers, the coaxial links enter what is known as a Hybrid-Fiber-Coax (HFC) network, digitally transporting the video and data traffic over a fiber infrastructure to the subscriber neighborhoods. Here, it is converted back to coax. Within the house, the coax splits to the TV and to the cable router. This cable router supports both PC and VoIP connectivity. A given downstream cable modem channel will initially service on the order of 15,000 households (assuming 1% cable modem service penetration, yielding 150 subscribers). As density grows, the serving area for this channel will be subdivided.

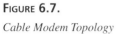

FIGURE **6.7.**

Cable Modem Topology

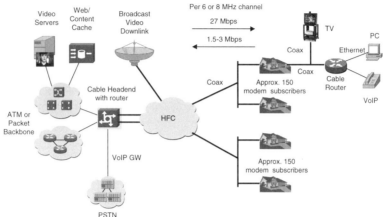

Subscribers contend for access to the data bandwidth, with the net result that cable modem installations operate as a giant hub. As more users access the service, they contend for the available bandwidth, and in times of peak usage, the system exhibits the characteristics of shared media such as Ethernet.

Support for real-time traffic such as VoIP therefore requires QoS awareness. An option here is to deploy RSVP between the CPE and CMTS or headend. The CMTS will assign a SID, as previously described, based on the RSVP request. DOCSIS 1.1 defines a number of service types that the subscriber may request, depending on the types of bandwidth guarantees required.

Parallel to implementing subscriber flow QoS, as the total volume of data traffic grows, the provider may decrease the number of subscribers served by a single frequency band as described above.

6.7 POTS/ISDN Dial-Up Options

Although not totally relevant to the broadband focus of this book, POTS and ISDN access has, in the past, met the needs of many of the subscribers in the same market segments addressed by ADSL—the residence, telecommuter, and small business segments. Analog dial relies on traditional modems, and has reached the theoretical (Shannon) capacity of the local loop with the availability in 1998 of V.90 modems operating at 56 Kbps. Even with the deployment of DSL and cable modems, the vast majority of Internet access from the residence will remain analog, and since almost all PCs and laptops are equipped with modems, this means access will remain significant for a long time to come.

ISDN, at least in the United States, is another story. After many false starts during the 1980s and early 1990s, by 1998, ISDN had gained a major foothold for Internet access—that is, until DSL and cable modems. ISDN is more costly on a monthly basis than either of these two technologies and delivers less bandwidth. Into 1999, new ISDN installations will slow as users transition to DSL and cable modems. The situation is a bit different in Europe, where ISDN has a major presence in some countries as both a data service and for voice (as in Germany). In these environments, and due to delays in deploying DSL and cable modems, ISDN should continue to experience strong growth.

6.8 T1, Wireless, and Satellite Services

Other alternatives to ADSL include the currently deployed T1/E1 and fractional services, although these do not meet the price points of the intended ADSL market. The same applies for Frame Relay, where current 256 Kbps tariffs are well above those of equivalent DSL services. Although these rates are expected to drop over time, the same may be said of ADSL.

At higher speeds, native ATM access will continue to remain a viable alternative, but once again, the intended audience is different. For all of these technologies, the provisioning and monitoring systems in place are designed to support on the order of thousands or tens of thousands of subscribers, unlike ADSL's subscriber base in the millions. This is possibly the greatest disconnect of all.

6.8.1 Wireless Technologies

One technology that holds promise is wireless, assuming the equipment and access costs are pushed lower than where they are today. Although high speed services such as the Local Multipoint Distribution Service (LMDS) and the Multichannel Multipoint Distribution System (MMDS) are first targeted at businesses, in some environments they may serve residential subscribers at megabit speeds as well.

As an example, Alcatel (1998) predicts that in the year 2003, 9 percent of broadband access will be via LMDS, compared to 26 percent for cable modems, 36 percent for DSL, 12 percent for ISDN, and 12 percent for satellite. This is good for an access technology where the infrastructure does not yet exist.

The question then is whether the cost of this infrastructure will ever compete with the installed copper (or cable) plant. In areas where neither exists, LMDS/MMDS will probably be very attractive, in the same way that cellular service is attractive for POTS traffic in those countries without sufficient copper infrastructure.

LMDS is a line-of-sight technology offering data rates of up to 155 Mbps downstream at distances to six miles. This 155 Mbps is contained within a 40 MHz channel located between the following frequencies:

27.5 and 28.35 GHz	To subscriber	pt-mpt
29.24 and 29.375 GHz	From subscriber	pt-pt
31.025 and 31.225 GHz	From subscriber	pt-pt

Note that frequencies in other countries may be different. A single hub site will serve between 1000 and 4000 subscribers depending on the data rates involved. As with cable, the upstream data rate is significantly lower, since subscribers contend for timeslots via a MAC protocol. This protocol allows a CPE to reserve upstream timeslots. These may be transient for bursty traffic or permanent for CBR-like traffic. The typical subscriber will have access to a T1/E1 worth of upstream bandwidth.

Downstream encoding is based on QPSK, while upstream data relies on DQPSK, both encoding techniques resilient to interference. This latter point is quite critical, since the 28 GHz band is susceptible to loss from rain. As an example, if the service provider wishes to offer a 99.999 percent service guarantee (5 minutes per year downtime), the maximum acceptable distance will span from .65 miles in New Orleans (rainy) to 1.4 miles in Phoenix (dry). By the same token, a 99.9 percent uptime (8.7 hours per year downtime) results in a 2.2 and 6.4 mile range for the same locations.

MMDS operates in a different frequency spectrum with more widely spaced hubs. Typically, hub sites serve a radius of 40–50 miles covering between 100 and 300 thousand subscribers. It operates in the 2.5 GHz band, with 198 MHz of available spectrum for the service. 2.5–2.668 GHz supports a pt-mpt downlink, while 2.15–2.162 GHz is the point-to-point uplink from the subscriber. A typical 6 MHz channel will support up to 27 mbps of data, almost identical to cable. This downstream bandwidth is QAM encoded, while upstream data relies on QPSK. As the LMDS, users contend for upstream bandwidth.

In a typical deployment (such as the one shown in Figure 6.8), the LMDS or MMDS hub station connects to an ATM backbone network. This then interfaces with any video and web caches, along with the digital broadcast video downlink. Each subscriber is equipped with an antenna and converter, with the broadcast signal split off to a TV and the data signal to a router. This router supports both PCs and VoIP.

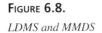

Figure 6.8.

LDMS and MMDS

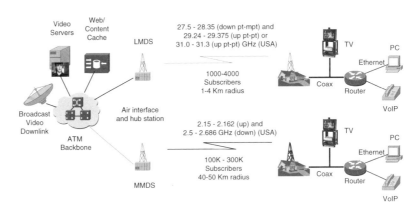

6.8.2 Rural Subscriber Issues

Note that none of these technologies, DSL included, really addresses the needs of the rural subscriber—subscribers beyond the reach of cable buildouts, current DLC systems, or where the subscriber density does not warrant MMDS. Here, satellite is probably the only option, and one promising system is the DirecPC which relies on high-bandwidth downlinks and telco return. This is obviously not the optimal solution, but it is a start.

LEO systems such as Teledesic, although capable of megabit data rates, will probably be priced out of the range of the residential subscriber for a long time to come. The next generation of cellular systems should begin to address the problem, realizing that existing systems are limited to about 9600 bps, but once again, tariffs may not encourage casual use. The real solution will be for the rural subscriber to wait for DLC buildouts, which are in fact occurring. Alternatively, as technology improves, some of the two-way satellite systems based on shared access are showing promise for the future.

Endnotes:

ADSL Forum, 3VDSL2 presentation, September 1998.

Alcatel, Presentation at Loop 198 conference, Santa Clara, CA, 1998.

Aware, 3wDSL Solution to Digital Carrier Service Provisioning,2 1998,
`http://www.aware.com/technology/whitepapers/wdsl.html`.

Yankee, Outlook(ing) for the ADSL Equipment Market, Data Communications, Report Vol 13 No 14, July, 1998.

Appendix **A**

ADSL Tariffs and NEBS Requirements

Table A.1 is a representative list of ADSL tariffs in the United States and elsewhere, current as of Spring 1999. Of course, the various service offerings will change over time, and the reader is therefore referred to the web sites referenced (which are also subject to change).

The table first lists the provider, and then the service name, if any, and the bandwidths available. In some cases, the bandwidth is guaranteed, while in others it is speficied as a maximum. A monthly tariff applies to a given bandwidth, with an ISP component either bundled into the base tariff or billed separately. This varies by provider.

The final column lists any additional fees. These usually include basic installation as well as the ADSL modem costs. As part of some promotional offerings, fees for installation, modem, or both are waived (many times requiring a service term commitment, much like some cellular offerings).

TABLE A.1. SAMPLE ADSL TARIFFS

Provider	Service	Down-stream/ Upstream BW	Monthly Fee	ISP Access (if additional)	Additional Fees
US West (US ILEC) www.uswest.com/com/ customers/interprise/dsl/ pricing.html	MegaLine MegaOffice MegaBusiness	256/256 kbps 512/512 kbps 768/768 kbps	$40 $62.40 $76.80	$19.95 $22.55 $23.15	Initial setup fee ($110) and modem ($199-$299) Modem fee waived if user-installed
	MegaBit	1/1 Mbps 4/1 Mbps 7/1 Mbps	$120 $480 $840	$19.95 $19.95 $19.95	
US West.net (Inprise) (US ISP) www.uswest.com/com/ customers/uswestnet/ service.html	Internet Access - DSL	–7/1 Mbps	$19.95-$139.95	Included	DSL fees; $25 setup
	OfficeWorks	–7/1 Mbps	$34.95-$154.95	Included	DSL fees; $25 setup
	OfficeWorks LAN	–7/1 Mbps	$79.95-$199.95	Included	DSL fees; $50 setup
	MegaPak MegaPak - Officeworks	256/256 kbps 512/512 kbps	$59.95	Included Included	Initial setup + modem + PC NIC ($500)
BellSouth (ILEC) + BellSouth.net (US ISP) www.bellsouth.net/ external/adsl/	FastAccess	1.5 Mbps/256 kbps	$59.95 $49.95 if Bell-South voice customer	Included	$199.95 modem; $99.95 installation
SBC - Pacific Bell (US ILEC) + Pacific Bell Internet Service public.pac-bell.net/dedicated/dsl	FasTrak DSL - Basic Home Enhanced (multiuser) Business	384-1.5/128 1.5-6/ 384 384-1.5/384 1.5-6 / 384 384-1.5 / 128 1.5-6 /384	$39 $129 $39 $129 $39 $129	$10 $30 $40 $70 $199 $199	$198 modem + NIC; installation waived with term agreement of 1 or 3 years $973 modem + router
Bell Atlantic + Bell Atlantic.net (US ILEC) www.bellatlantic.com/ infospeed/	Infospeed DSL – Personal Professional Power	640 /90 1.6/90 7.1/680	$39.95 $59.95 $109.95	$20 $50 $80	$49 modem w/ term agreement $99 service connection Inst $99 and modem $325 w/o term
Cincinnati Bell (US ILEC) www.cincinnati-bell.com/adsl	ZoomTown - BaseSpeed TurboSpeed HyperSpeed	384 / 90 768 / 384 1.5 / 768	$29.95 $59.95 $159.95	Included	Installation + equip waived if 12 month term; $350 w/o term
GTE (US ILEC) www.gte.com/dsl/idsl/ dslprice.html	DSL Bronze DSL Silver DSL Gold DSL Platinum DSL Platinum Plus (multiuser)	256/64 384/384 768/768 1.5/768 1.5/768	$35 $55 $70 $100 $230	Additional	$60 installation $300 modem
Telus (Canada) www.telusplanet.net/ hispeed/pricede-tails.html	PLAnet EXTREME	2.5 / 1	$49.95 (volume limited to 5 GB / 200 MB; single user)		$150 installation
	PLAnet BizExpress	2.5 / 1	$99.95 (multiple users; dedicated; 10 GB / 2 GB per month)		$175 installation
Brainstorm www.brainstorm.net (US ISP)	Via Covad Via NorthPoint	1.5 / 384 1.5	$370 / 1 Gbit $425 / 1 Gbit $5 / addl. Gbit	Included Included	$499 installation $499 installation

Table A.2 lists the various Network Equipment Building System requirements relevant to DSLAM (and for that matter, any) Central Office installations. The first column lists the general category, the second the actual document, and the third the reasoning for the requirement.

There are three levels of compliance: Level 1, 2, and 3. Thus, when a vendor claims that it has passed "NEBS Level 3," it has demonstrated compliance with all levels. For more information on the NEBS requirements listed in this table, go to Bellcore at `http://www.bellcore.com/BC.dynjava?NEBSTestingCHPGeneralContent HomePage`.

TABLE A.2. NETWORK EQUIPMENT-BUILDING SYSTEM (NEBS) REQUIREMENTS

NEBS Level 1 Requirements

System Fire Test and Material/Components Criteria	GR-63 Criteria 80–103	Ensure that device is not a fire hazard.
Electrical Safety	GR-1089 Criteria 54–71	Ensure that voltage/current sources are properly protected from accidental contact. Ensure that leakage current is acceptable.
Listing Requirements	GR-1089 Criteria 22–24	Applies to AC powered equipment at the customer premise.
Bonding and Grounding	GR-1089 Criteria 74–93	Ensure that system can be properly grounded to the CO GRD and that the internal DC-DC converters can handle a short circuit condition.
EMI Emissions	GR-1089 Criteria 8-14	Ensure that system meets acceptable levels of conducted and radiated emissions.
Short-Circuit Test	GR-1089 Criterion 25	Ensure that no fire hazard results from short-circuiting the tip and ring connections to each other and to ground.
Lightning Immunity, Second Level	GR-1089 Criteria 29 and 33	Ensure that the system does not become a fire or safety hazard as a result of application of a simulated lightning surge across tip and ring. Criteria applies to AC systems.

A

TABLE A.2.   NETWORK EQUIPMENT-BUILDING SYSTEM (NEBS) REQUIREMENTS (*CONTINUED*)

AC Power Fault Immunity, Second Level	GR-1089 Criteria 36–41	Ensure that the system does not become a fire or safety hazard when subjected to an AC voltage across tip and ring or from tip/ring to ground. Testing includes both loop and CO side connections.
Current Limiting Protector Test	GR-1089 Criteria 21 and 34	Describes additional protection devices that may be necessary in the event of failure of the lightning/power cross tests.
Voltage Limiting Protector Test	GR-1089 Criterion 20	Describes requirements on additional protection devices that may be necessary in the event of failure of the lightning/ power cross tests.
NEBS Level 2 Requirements		
Operational Thermal, Operating Conditions	GR-63 Criterion 72	Ability of system to operate continuously in the temperature range of +5°C to +40°C and a relative humidity of 10–85%.
Earthquake Zone 2	GR-63 Criteria 110-112, 114, 115, 117, 119	Ability of system to withstand earthquakes at a Zone 2 level.
Office Vibrations	GR-63 Criteria 122, 123	Ability of system to withstand and operate during low level vibrations along three perpendicular axes.
Airborne Contaminants, Indoor levels	GR-63 Criterion 125	Ability of system to resist damage when subjected to pollutants. Test is for corrosion and dust.
ESD Normal Operation, Installation, and Repair	GR-1089 Criteria 1–3, 5–7	Ability of system to resist various electrostatic discharges with/without wrist straps.
EMI-Emissions	GR-1089 Criteria 8, 10, 11, 13, 14	Ability of system to meet acceptable levels of conducted and radiated emissions beyond Level 1.
EMI-Immunity	GR-1089 Criteria 15, 17, 19	Ability of system to withstand radiated and conducted noise.
Lightning Immunity, First Level	GR-1089 Criteria 27, 28, 30-32	Ability of system to remain operational during and after application of a simulated lightning surge across tip and ring.

TABLE A.2. NETWORK EQUIPMENT-BUILDING SYSTEM (NEBS) REQUIREMENTS *(CONTINUED)*

AC Power Fault Immunity, First Level	GR-1089 Criterion 35	Ability of system to remain operational during and after application of an AC power fault across tip and ring or from tip/ring to ground. AC fault may be as large as 600 volts for as long as 15 minutes.
Steady State Power Induction Requirements	GR-1089 Criteria 42, 44, 46-49, 51	Longitudinal induction for DSL equipment.
NEBS Level 3 Requirements		
Operational Thermal, Short-term Conditions	GR-63 Criterion 73	Ability of system to operate in the temperature range of –5°C to 50°C and a relative humidity range of 10–90% under short term conditions (16 hours maximum).
Earthquake Zone 4	GR-63 Criteria 110-112, 114-115, 117, 119	Ability of system to withstand earthquakes at a Zone 4 level.
Airborne Contaminants Outdoor Levels	GR-63 Criteria 126, 127	Ability of system to operate in outside enclosures.
ESD Installation and Repair	GR-1089 Criterion 4	Ability of system to withstand various electrostatic discharges to surfaces contacted only during installation and repair.
EMI Emissions	GR-1089 Criterion 9	Addresses emissions with doors closed versus open.
EMI Immunity	GR-1089 Criteria 16, 18	Ability of system to withstand higher levels of radiated electric fields than with Level 2.
Steady-State Power Induction Conditional Requirements	GR-1089 Criteria 43, 45, 50, 52	Does not apply to DSL equipment.
Additional NEBS Requirements		
Spatial Requirements	GR-63 Criteria 1–42	
Altitude	GR-63 Criteria 74, 76	
Thermal Heat Dissipation	GR-63 Criteria 77–79	
Acoustic Noise	GR-63 Criterion 128	
Illumination	GR-63 Criteria 129–135	
DC Potential Difference	GR-1089 Criterion 53	

A

Appendix B

Relevant Documents, Standards, and Resources

This appendix lists the various standards relevant in the ADSL space. These include those defining protocols and architectures from the ADSL Forum, the basic protocol documents (RFCs and drafts) from the IETF, ATM-specific standards from the ATM Forum, and standards relating to a number of areas from the ITU-T.

The next tables list the vendors active in the DSL space, as well as a representative list of global service providers offering or soon to offer ADSL service. Finally, a brief list of on-line resources in the ADSL space is included. Note that all of these references are current as of Spring 1999 and subject to change over time. Thus, the reader is directed to the web sites of the various standardization bodies, vendors, and service providers for the most current information.

Relevant ADSL Forum Documents

For more information, see the ADSL Forum at www.adsl.com.

Document	Name	Date
PR-001	ADSL Recommendation Number 1 – Splitter mode operation	December, 1998
PR-002	ADSL Recommendation Number 2 – Splitterless operation	December, 1998
PR-003	ADSL Recommendation Number 3 – ATM compliance	December, 1998
TR-001	ADSL Forum System Reference Model	May, 1996
TR-002	ATM over ADSL Recommendations	March, 1997
TR-003	Framing and Encapsulation Standards for ADSL: Packet Mode	June, 1997
TR-004	Network Migration	December, 1997
TR-005	ADSL Network Element Management	March, 1998
TR-006	SNMP-based ADSL Line MIB	March, 1998
TR-007	Interfaces and System Configurations for ADSL: Customer Premises	March, 1998
TR-008	Default VPI/VCI Addresses for FUNI Mode Transport: Packet Mode	March, 1998
TR-009	Channelization for DMT and CAP ADSL Line Codes: Packet Mode	March, 1998
TR-010	Requirements & Reference Models for ADSL Access Networks: The "SNAG" Document	June, 1998
TR-011	An End-To-End Packet Mode Architecture With Tunneling And Service Selection	June, 1998
TR-012	Broadband Service Architecture for Access to Legacy Data Networks over ADSL Issue 1	June, 1998
TR-013	Interface and Configurations for ADSL	January, 1999
TR-014	DMT Line Code Specific MIB	December, 1998
TR-015	CAP Line Code Specific MIB	February, 1999
TR-016	CMIP Specification for ADSL Network Element Management	January, 1999
WT-018	Interfaces and System Configurations for ADSL: Central Office	
WT-021	ATM Mode Working Text for TR-002 Version 2	
WT-024	ADSL Network Management Architecture	
WT-026	The Operation of ADSL-based Networks	
WT-027	Overview of ADSL Testing	
WT-028	ADSL Conformance Testing	
WT-029	ADSL Performance Measurements	
WT-030	ADSL Basic Interoperability specification	
WT-031	References and Requirements for CPE Architectures for Data Access	
WT-032	Premises Architectures	
WT-033	Core Network Architecture for Access to Legacy Data Networks over ADSL	

Document	Name	Date
WT-034	T1.314 Issue2 ADSL ICS (Interface Conformance Statement)	
WT-035	ITU-T G.992.2 ICS	

Note: Working Text numbering not necessarily contiguous as documents are approved and become Technical Reports.
PR – Proposed Recommendation
TR – Technical Report
WT – Working Test

B

Relevant IETF Documents

For more information, see the Internet Engineering Task Force at www.ietf.org.

RFC	Title*	Description
791	Internet Protocol	IP Standard
793	Transmission Control Protocol	TCP Standard
768	User Datagram Protocol	UDP Standard
826	Address Resolution Protocol	MAC to IP address resolution
1034	Domain Names - Concepts and Facilities Host extensions for IP multicasting	Description of the DNS Outlines IP multicasting
1142	OSI IS-IS Intra-domain Routing Protocol	
1155	SMI for TCP/IP	Structure of Management Information for SNMPv1
1213	Management Information Base for Network	For SNMPv1 Management of TCP/IP-based Internets: MIB-II
1332	The PPP Internet Protocol Control Protocol (IPCP)	Control for PPP
1483	Multiprotocol Encapsulation over ATM Adaptation Layer 5, 1993	Describes encapsulations for carrying network interconnect traffic over ATM AAL5
1631	The IP Network Address Translator (NAT)	
1662	PPP in HDLC-like Framing.	PPP over frame connections
1723	RIP Version 2 - Carrying Additional Information	Intra-domain routing
1771	A Border Gateway Protocol 4 (BGP-4)	Interdomain routing
1883	Internet Protocol, Version 6 (IPv6) Specification	Next version of IP
1901	Community-based SNMPv2	Simplified SNMPv2
1918	Address Allocation for Private Internets	Private address spaces

RFC	Title*	Description
1932	IP over ATM: A Framework Document Dynamic Host Configuration Protocol (DHCP)	Describes the various proposals of mapping IP into ATM.
		Automated IP address assignment
2138	Remote Authentication Dial In User Service (RADIUS)	User authentication and authorization
2205	Resource Reservation Protocol (RSVP)	Application QoS Support
2212	Specification of Guaranteed Quality of Service Specification of Controlled Load Quality of Service	Describes the network element behavior required to deliver guaranteed service in the Internet as above for Controlled Loa
2225	Classical IP and ARP over ATM Update	Support for SVCs over ATM. Updates RFCs 1577 and 1626
2251	Lightweight Directory Access Protocol (v3) Architecture for describing SNMP management frameworks	User information SNMPv3 management
2272	Message processing and dispatching for SNMPv3	Management
2273	SNMPv3 applications	Management
	User-based Security Model for SNMPv3	Management
2328	OSPF Version 2	Intra-domain routing
	ATM Signaling Support for IP over ATM –	
	UNI Signaling 4.0 Update	Describes support for IP over ATM signaling. Replaces RFC 1755
2362	Protocol Independent Multicasting (PIM) Sparse-mode	Multicast routing
	Protocol Independent Multicasting (PIM) Dense-mode	Multicast routing draft-ietf-pim-v2-dm-01.txt
2364	PPP over AAL5	PPP over ATM
2401	Security Architecture for the Internet Protocol	
2402	IP Authentication Header	For IPSec
2406	IP Encapsulating Security Payload (ESP)	For IPSec
2408	Internet Security Association and Key Management Protocol (ISAKMP)	For IPSec
2409	Internet Key Exchange (IKE)	For IPSec
2427	Multiprotocol Interconnect over Frame Relay	Updates RFC 1490
2516	PPP over Ethernet	Transports PPP over Ethernet

RFC	Title*	Description
	The TACACS+ Protocol Version 1.78 AAA	draft-grant-tacacs-0x.txt Authentication
	A Framework for Multiprotocol Label Switching	draft-ietf-mpls-framework-0x.txt. Efficient IP forwarding
	Layer Two Tunneling Protocol - L2TP	draft-ietf-pppext-l2tp-1x.txt Secure tunneling
	PGM Reliable Transport Protocol Specification	draft-speakman-pgm-spec-0x.txt IP multicasting
	Assured Forwarding PHB Group	draft-ietf-diffserv-af-0x.txt For Diffserv
	Expedited Forwarding PHB	draft-ietf-diffserv-phb-ef-0x.txt For Diffserv

B

Relevant ATM Forum Documents

For more information, see the ATM Forum at www.atmforum.com.

Document Title	Date	Document Number
LAN Emulation over ATM 1.0	1995	af-lane-0021.000
LANE v2.0 LUNI Interface	1997	af-lane-0084.000
Multi-Protocol over ATM v1.0	1997	af-mpoa-0087.000
P-NNI V1.0	1996	af-pnni-0055.000
PNNI Addendum – ABR Support	1997	af-pnni-0075.000
PNNI Addendums – Mobility, Security	(1999)	
FUNI 2.0	1997	af-saa-0088.000
Circuit Emulation 2.0	1996	af-vtoa-0078.000
VTOA to the Desktop	1997	af-vtoa-0083.000
Dynamic Bandwidth Utilization	1997	af-vtoa-0085.000
ATM Trunking for Narrowband Services	1997	af-vtoa-0089.000
ATM User-Network Interface Specification V3.1	1994	af-uni-0010.002
UNI Signaling 4.0	1996	af-sig-0061.000
ILMI v4.0	1996	af-ilmi-0065.000

Document Title	Date	Document Number
Traffic Management 4.0	1996	af-tm-0056.000
Signaling ABR Addendum	1997	af-sig-0076.000
Traffic Management ABR Addendum	1997	af-tm-0077.000
Traffic Management 4.1	(1999)	
RBB Architecture Framework	1998	af-rbb-0099.000
Security Framework 1.0	1998	af-sec-0096.000

Dates in parenthesis are projected.

Relevant ITU-T Standards

For more information, see the International Telecommunications Union at www.itu.int.

Number	Document Title and Date
I.361	BISDN ATM Layer Specification, November, 1995 (A more detailed description of the information in I.150.)
I.362	BISDN Asynchronous Transfer Model Adaptation Layer, October, 1993 (The principles of the AALs corresponding to the BISDN service classes.)
I.363	Asynchronous Transfer Mode Adaptation Layer Specification, February, 1994/1996 (A more detailed description of the AALs from I.362, including the division of the AAL into the Segmentation and Reassembly sublayer and the Convergence Sublayer.)
I.363.1	BISDN ATM Adaptation Layer (AAL) Specification, Type 1, August, 1996
I.363.5	BISDN ATM Adaptation Layer (AAL) Specification, Type 5, August, 1996
G.991.1	High Bitrate Digital Subscriber Line (HDSL) Transmission System on Metallic Local Lines, Draft ITU-T Recommendation G.991.1, October, 1998.
G.992.1	Asymmetrical Digital Subscriber Line (ADSL) Transceivers, Draft ITU-T Recommendation G.992.1 (ex: G.dmt), October, 1998.
G.992.2	Splitterless Asymmetric Digital Subscriber Line (ADSL) Transceivers, Draft Recommendation G.992.2, October, 1998.
G.994.1	Handshake procedures for Digital Subscriber Line (DSL) transceivers, Draft ITU-T Recommendation G.994.1, October, 1998.
G.995.1	Overview of Digital Subscriber Line (DSL) Recommendations, Draft Recommendation G.995.1, October, 1998.
G.996.1	Test Procedures for Digital Subscriber Line (DSL) Transceivers, Draft ITU-T Recommendation G.996.1, October, 1998.

B

Number	Document Title and Date
G.997.1	Physical Layer Management for Digital Subscriber Line (DSL) Transceivers. Draft ITU-T Recommendation G.997.1, October, 1998.
G.VB51	V-interfaces at the Service Node (SN) - VB5.1 Reference Point Specification, September, 1997 (draft)
G.VB52	V-interfaces at the Service Node (SN) - VB5.2 Reference Point Specification, March, 1998 (draft)
G.ATMA	Architecture of Transport Networks based on ATM
H.320	Narrow-band Visual Telephone Systems and Terminal Equipment
H.321	Visual Telephone Terminals over ATM
H.323	Visual Telephone Terminals over Non-Guaranteed Quality of Service LANs
Q.2010	B-ISDN overview. Signaling Capability Set 1, Release 1, August, 1991
Q.2100	B-ISDN Signaling ATM Adaptation Layer (SAAL) Overview Description, February, 1995
Q.2110	B-ISDN ATM Adaptation Layer Service Specific Connection Oriented Protocol (SSCOP), April, 1995
Q.2119	B-ISDNATM Adaptation Layer Protocols – Convergence Function for the SSCOP above the Frame Relay Core Service (1996)
Q.2120	B-ISDN meta-signaling protocol, September, 1995
Q.2130	B-ISDN SAAL Service Specific Coordination Function (SSCF), April, 1995
Q.2140	B-ISDN ATM adaptation layer – Service-specific coordination function for signaling at the network–node interface (SSCF at NNI), November, 1995
Q.2144	B-ISDN Signaling ATM Adaptation Layer (SAAL) – Layer Management for the SAAL at the Network Node Interface, December, 1995
Q.2210	B-ISDN Signaling Network protocols. Message Transfer Part level 3 functions and messages using the services of ITU-T Recommendation Q.2140 (1996)
Q.2931	B-ISDN Digital Subscriber Signaling System No. 2 (DSS2) User–Network Interface (UNI) Layer 3 Specification for Basic Call/Connection control (modified by Q.2971), February 1995
Q.2971	B-ISDN DSS2 UNI Layer 3 Specification for Point-to-Multipoint Call/Connection Control (modifies Q.2931, Q.2951, and Q.2957), December, 1995

ADSL Vendors and Codes

Vendor	Web Site	HW vendor code (if assigned)
Accelerated Access	www.acceleratedaccess.com	
ADC Telecommunications	www.adc.com	0014
Adtran	www.adtran.com	0012
AG Communications Systems	www.ag.com	002F
Alcatel Network Systems	www.alcatel.com	0022
Amati Communications	www.amati.com	0006
AMD	www.amd.com	0041
Analog Devices	www.analog.com	001C
Ariel	www.ariel.com	002A
Ascend	www.ascend.com	
Assured Access (part of Alcatel)	www.assuredaccess.com	
AWA	www.awa.com	0021
Aware	www.aware.com	000B
Brooktree	www.brooktree.com	000C
Cayman Systems	www.cayman.com	0048
Cisco Systems	www.cisco.com	0046/0034
Compaq Computer	www.compaq.com	003C
Copper Mountain	www.coppermountain.com	0036
Cosine Networks	www.cosinecom.com	
Diamond Lane (part of Nokia)	www.diamondlane.com	0029
DSC (part of Alcatel)	www.dsc.com	0018
ECI Telecom	www.eci.com	0003
Efficient Networks	www.efficient.com	0044
Flowpoint (part of Cabletron)	www.flowpoint.com	0049
Fujitsu Network Transmission Systems	www.fujitsu.com	0026
Globespan	www.globespan.com	0039
Harris	www.harris.com	0031
IBM	www.ibm.com	0016
INC	www.inc.com	0013

Vendor	Web Site	HW vendor code (if assigned)
Interspeed	www.interspeed.com	0045
Intel	www.intel.com	0005
Level One Communications	www.level1.com	0008
Lucent Technologies	www.lucent.com	000A
Metalink	www.metalink.com	002C
Multi-Tech Systems	www.multitech.com	0040
Motorola	www.mot.com	0015
National Semiconductor	www.national.com	0023
NEC	www.nec.com	000D
Newbridge Networks	www.newbridge.com	0017
Next Level Communications	www.nextlevel.com	003F
Nokia	www.nokia.com	001D
Nortel Networks (inc. Bay Networks)	www.nortelnetworks.com	000F/old bay one
Northchurch Communications	www.northchurchcom.com	
Orckit Communications	www.orckit.com	0020
PairGain Technologies	www.pairgain.com	0010
Paradyne	www.paradyne.com	0011
Promatory Communications	www.promatory.com	
Pulsecom	www.pulsecom.com	002D
RedBack	www.redback.com	
RedStone (part of Siemens)	www.redstonecom.com	
Rockwell	www.rockwell.com	0030
RouterWare	www.routerware.com	
Samsung	www.samsung.com	000E
Shasta Networks (part of Nortel Networks)	www.shastanets.com	
Siemens Telecom	www.siemens.com	001B
Spring Tide Networks	www.springtidenet.com	
Sumitomo Electric	www.sumitomo.com	0042
Tellabs Operations	www.tellabs.com	001F
Texas Instruments	www.ti.com	0004
Tollgrade Communications	www.tollgrade.com	0047
Turnstone Systems	www.turnstonesystems.com	

B

Vendor	Web Site	HW vendor code (if assigned)
Westell	www.westell.com	0002
3Com	www.3com.com	0035/002E

Note: The hardware vendor code is used by management systems to identify the source of the deployed hardware.

Representative List of Service Providers

Service Provider	Type	Web site
Ameritech (SBC)	US ILEC / ISP	www.ameritech.com (net)
Bell Atlantic	US ILEC / ISP	www.bellatlantic.com (net)
BellSouth	US ILEC / ISP	www.bellsouth.com (net)
British Telecom	UK PTT / ISP	www.britishtelecom.com (net)
Cable and Wireless	UK IXC / ISP	www.cablewireless.com
Cincinnati Bell	US ILEC / ISP	www.cincinnatibell.com
Concentric Network	US ISP	www.concentric.net
Covad Communications	US PCLEC	www.covad.com
Demon Internet	UK ISP	www.demon.net
Deutsche Telecom (Germany)	Germany PTT / ISP	www.dtag.de
Exodus	US ISP	www.exodus.com
France Telecom	France PTT / ISP	www.francetelecom.com (net)
Frontier GlobalCenter	US CLEC	www.globalcenter.com
GTE	US ILEC / ISP	www.gte.com (net)
Level3	US CLEC	www.level3.com
MCI Worldcom	US IXC/ISP	www.mciworldcom.com
NorthPoint	US PCLEC	www.northpointcom.com
NTT	Japan PTT	www.ntt.co.jp
Quest Communications	US CLEC	www.qwest.com
Pacific Bell (SBC)	US ILEC / ISP	www.pacbell.com (net)
Rhythms NetConnections	US CLEC	www.rhythms.net
Rocky Mountain Internet	US ISP	www.rmi.com
Southwestern Bell (SBC)	US ILEC / ISP	www.swbell.com (net)

Service Provider	Type	Web site
Sonera	Finland PTT / ISP	www.sonera.fi
Telecom Italia	Italy PTT	www.telecomitalia.it
Telefonica	Spain PTT	www.telefonica.es
Telia	Sweden PTT / ISP	www.telia.se (net)
Telstra	Australia PTT / ISP	www.telstra.com.au
US West	US ILEC / ISP	www.uswest.com (net)
UUNET	US ISP	www.uu.net
Verio	US ISP	www.verio.com

Provider types: ILEC – Incumbent Local Exchange Carrier

CLEC – Competitive Local Exchange Carrier

PCLEC – Packet CLEC

IXC – Interexchange Carrier (Long Distance)

ISP – Internet Service Provider

PTT – Post, Telephone, and Telegraph

In the website column, 'net' implies that the ISP component of the listed service provider is reachable over www.provider-name.net

On-Line Sources of Information

Source	Web site
CMP Media	www.cmp.com
Network World	www.nwfusion.com
Telechoice DSL Site	www.xdsl.com
Nortel Networks CLEC Site	www.nortel.com/pcn/isp/goals/beclec.htm
General information on CLECs	www.clec.com

Many vendors maintain a whitepaper section within their web sites. Some of the better ones are at Aware, Analog Devices, Paradyne, and Nortel.

Recommended Reading

ATM:

- De Prycker, M., 'Asynchronous Transfer Mode, Solution for Broadband ISDN, 2ed', Ellis Horwood, 1993
- Ginsburg, D., 'ATM: Solutions for Enterprise Internetworking, 2ed', Addison Wesley, 1998.
- Black, U., 'ATM Resource Library, Volumes 1, 2, 3, 2ed', Prentice Hall, 1998.

DSL:

- Goralski, W., 'ADSL And DSL Technologies', McGraw-Hill, 1998.
- Chen, W., 'DSL SimulationTechniques and Standards Development For Digital Subscriber Line Systems', Macmillan Technical Publishing, 1998.

Security:

- Schneier, B., 'Applied Cryptography : Protocols, Algorithms And Source Code In C, 2ed', John Wiley & Sons, 1995.
- Cheswick, W., and Bellovin, S., 'Firewalls and Internet Security', Addison Wesley, 1994.

TCP/IP and Routing:

- Huitema, C., 'Routing In The Internet', Prentice Hall, 1995.
- Halabi, B., 'Internet Routing Architectures', Cisco Press, 1997
- Comer, D., 'Internetworking with TCP / IP, Volume 1: Principles, Protocols, and Architecture', Prentice Hall, 1995.

Appendix C

Glossary

Access Network – Portion of the network connecting Access Nodes to the subscribers. Consists of copper pairs and coaxial cable in most cases, but may be fiber or wireless.

Access Nodes – When individual access lines are concentrated into higher speed uplinks; DLC systems and ONUs are examples of access nodes.

ADSL (Asymmetric Digital Subscriber Line) – Technology offering asymmetric and symmetric bandwidth of up to 8 Mbps downstream over two copper pairs and supporting POTS splitters; encoding based on DMT standardized by ANSI and the ITU-T.

ADSL Lite – Variant of ADSL with a maximum downstream bandwidth of 1.5 Mbps and not requiring splitters; intended for consumer market; see also UAWG.

ANSI (American National Standards Institute) – US telecommunications standards setting body. European equivalent is *ETSI*.

ATM (Asynchronous Transfer Mode) – Technology segmenting all voice, video, and data traffic into fixed length *cells*; preferred encapsulation of ADSL loops.

ATU-C/ATU-R (ADSL Terminating Unit – Central/Remote) – The ADSL modems at the CO and at the subscriber; the ATU-C is usually within the DSLAM.

Bridge Taps - Sections of un-terminated twisted-pairs connected in parallel across copper pairs; impacts range.

CAP (Carrierless Amplitude/Phase) – ADSL encoding technique spreading the data signal over a single carrier occupying the entire copper frequency spectrum; not standardized.

CLEC (Competitive Local Exchange Carrier) – Service provider created to offer value-added voice and data services in competition with the ILEC; formalized as part of the 1996 Telecommunications Act in the US.

CPE (Customer Premises Equipment) – Networking hardware located at the subscriber's site.

CO (Central Office) – Building originally containing only voice switches and owned by the ILEC; CLECs may place hardware within the CO.

Co-location – Describes a CLEC placing networking equipment within the premises of an ILEC, usually in the CO.

CSA (Carrier Serving Area) – Area served by a given CO.

DMT (Discrete Multitone) – ADSL encoding technique using subchannels and standardized by ANSI and the ITU-T; some ADSL installations are also based on CAP.

DLC (Digital Loop Carrier) – Enclosure placed closer to the subscriber to extend the reach of digital services. It is connected to the CO via fiber or copper.

DSL (Digital Subscriber Line) – Refers to any of the DSL technologies, including ADSL, HDSL, IDSL, SDSL, and VDSL. Has replaced xDSL in conversation.

DSLAM (Digital Subscriber Line Access Multiplexer) – Chassis containing DSL modems, aggregating them onto an ATM or frame uplink.

DS0 (Digital Signal 0) – Single 64Kbps voice or data channel; a DS1 contains 24 DS0s.

DS1 (Digital Signal 1) – Payload and framing structure for the North American 1.544 Mbps; used interchangeably with T1; other levels include DS3 at 45 Mbps.

E1 – European basic multiplex rate at 2.048 Mbps offering 30 times 64 Kbps channels; other levels include E3 at 34 Mbps.

ETSI (European Telecommunications Standardization Institute) – European standards organization; US equivalent is ANSI.

FEXT (Far-End Crosstalk) – Interference between two copper pairs at the subscriber end of the loop.

FTTH (Fiber to the Home) – Fiber extends from the CO directly to the subscriber's home.

FTTC (Fiber to the Curb) – Fiber extends from the CO to a curbside distribution point where the signal is converted to coaxial cable or twisted pair.

FTTN (Fiber to the Node) – Fiber extends from the CO to an intermediate point about 3000 feet from the subscriber; supports VDSL.

HFC (Hybrid Fiber-Coax) – Architecture within cable TV distribution where fiber is run to a neighborhood distribution point. The signal is then converted to coaxial cable.

HDSL (High-Speed Digital Subscriber Line) – Symmetric technology offering up to T1/E1 over two or four pairs. It does not support POTS splitters; overlaps somewhat with SDSL.

IDSL (Integrated Digital Subscriber Line) – Technology offering 144 Kbps over two copper pairs using ISDN's 2B1Q encoding; does not support POTS splitters.

ILEC (Incumbent Local Exchange Carrier) – The original telephony provider within a geographic area; the RBOC.

IAP (Internet Access Provider) – Offers basic Internet connectivity, does not necessarily own any facilities.

ISP (Internet Service Provider) – Entity offering Internet connectivity and value-added services; owns infrastructure.

IXC (Interexchange Carrier) – Service provider in the US offering long-distance telephony services.

LEC (Local Exchange Carrier) – US access and/or service provider resulting from the Telecommunications Act of 1996; see also ILEC and CLEC.

Loading Coils – Used on long copper loops to increase allowable voice distance; incompatible with high-speed digital services such as ADSL.

NAP (Network Access Provider) – Offers telephony services in the US.

NEXT (Near-End Crosstalk) – Interference between copper pairs at the CO end of the loop.

NSP (Network Service Provider) – Entity offering value-added services and connectivity within a telecommunications network; ISPs, ILECs, and CLECs are examples of NSPs.

NTE (Network Termination Equipment) – Terminates each end of the transmission line.

OC3 (Optical Carrier 3) – 155 Mbps transmission rate within North American SONET hierarchy; see also STS-3c.

ONU (Optical Network Unit) – Access node to convert optical signals to electrical signals for transmission over twisted pair or coaxial cable; part of FSAN deployments.

PCLEC (Packet CLEC) – Competitive provider offering only data services.

PON (Passive Optical Network) – Fiber-based transmission network with no active electronics; part of FSAN deployments.

POTS (Plain Old Telephone Service) – Refers to basic telephony service occupying the lowest 4 kHz of the copper frequency spectrum. ADSL supports POTS by shifting the data above the POTS frequencies.

POTS Splitter – Uses a filter to separate the telephony traffic from the DSL data; ADSL modems and DSLAMs are usually installed in conjunction with (or include) POTS splitters.

PTT (Post Telephone and Telegraph) – Used in Europe to refer to the former state-owned telephone companies.

RADSL (Rate Adaptive Digital Subscriber Line) – Refers to ADSL's ability to rate adapt to different line conditions; used infrequently.

RBOC (Regional Bell Operating Company) - Regional carriers resulting from the AT&T divestiture; also known as ILECs.

SDH (Synchronous Digital Hierarchy) – International optical hierarchy offering different multiplexing rates and including resilience; North American equivalent is SONET.

SDSL (Symmetric Digital Subscriber Line) - Offers symmetric service over two or four copper pairs and may support POTS; uses 2B1Q encoding but not standardized.

SONET (Synchronous Optical Network) - North American optical hierarchy offering different multiplexing rates and including resilience; international equivalent is SDH.

Splitterless ADSL - Also known as G.Lite/ADSL Lite; requires no in-home POTS filter installation simplifying installation.

STM-1 (Synchronous Transport Module 1) – Basic transport rate and framing of 155 Mbps within the SDH hierarchy; other levels include STM-4 at 622 Mbps and STM-16 at 2.4 Gbps.

STS-1 (Synchronous Transport Signal 1) - SONET basic transport rate of 51.84 Mbps; others include STS-3c (for concatenated) at 155 Mbps and STS-12c at 622 Mbps.

T1 - US basic multiplex rate at 1.544 Mbps offering 24 x 64 Kbps (DS0) channels.

Tier 1 – Refers to backbone or transit ISPs; includes Worldcom, GTE, AT&T, Quest, Level 3, and so on; regional and local ISPs are sometimes referred to as Tiers 2 and 3.

UAWG (Universal ADSL Working Group) - Consortium of PC, networking, and telecommunications vendors dedicated to promoting G.Lite/ADSL Lite, a simplified version of ADSL.

VDSL (Very High-Speed Digital Subscriber Line) - Asymmetric or symmetric technology based on a variation of **CAP/QAM or** DMT encoding delivering up to 52 Mbps over shorter distances (4500 ft maximum) than ADSL.

VOD (Video on Demand) - Refers to a service allowing subscribers to request video content at any time; ADSL offers the bandwidth required for VoD; variation is Near-VoD or staggercasting where content is streamed at set intervals (such as a popular movie every 15 minutes).

xDSL - Term used to refer to the various DSL technologies; more or less replaced by DSL.

C

Index

Index

Index

Addison-Wesley Computer and Engineering Publishing Group

How to Interact with Us

1. Visit our Web site

http://www.awl.com/cseng

When you think you've read enough, there's always more content for you at Addison-Wesley's web site. Our web site contains a directory of complete product information including:

- Chapters
- Exclusive author interviews
- Links to authors' pages
- Tables of contents
- Source code

You can also discover what tradeshows and conferences Addison-Wesley will be attending, read what others are saying about our titles, and find out where and when you can meet our authors and have them sign your book.

2. Subscribe to Our Email Mailing Lists

Subscribe to our electronic mailing lists and be the first to know when new books are publishing. Here's how it works: Sign up for our electronic mailing at http://www.awl.com/cseng/mailinglists.html. Just select the subject areas that interest you and you will receive notification via email when we publish a book in that area.

3. Contact Us via Email

cepubprof@awl.com

Ask general questions about our books.
Sign up for our electronic mailing lists.
Submit corrections for our web site.

bexpress@awl.com

Request an Addison-Wesley catalog.
Get answers to questions regarding your order or our products.

innovations@awl.com

Request a current Innovations Newsletter.

webmaster@awl.com

Send comments about our web site.

jcs@awl.com

Submit a book proposal.
Send errata for an Addison-Wesley book.

cepubpublicity@awl.com

Request a review copy for a member of the media interested in reviewing new Addison-Wesley titles.

We encourage you to patronize the many fine retailers who stock Addison-Wesley titles. Visit our online directory to find stores near you or visit our online store: http://store.awl.com/ or call 800-824-7799.

Addison Wesley Longman
Computer and Engineering Publishing Group
One Jacob Way, Reading, Massachusetts 01867 USA
EL 781-944-3700 • FAX 781-942-3076